Austin Flint

On the Source of Muscular Power

Austin Flint

On the Source of Muscular Power

ISBN/EAN: 9783337365721

Printed in Europe, USA, Canada, Australia, Japan

Cover: Foto ©berggeist007 / pixelio.de

More available books at **www.hansebooks.com**

ON

THE SOURCE

OF

MUSCULAR POWER.

ARGUMENTS AND CONCLUSIONS

DRAWN FROM

OBSERVATIONS UPON THE HUMAN SUBJECT, UNDER CONDITIONS OF REST AND OF MUSCULAR EXERCISE.

BY

AUSTIN FLINT, Jr., M. D.,

PROFESSOR OF PHYSIOLOGY AND PHYSIOLOGICAL ANATOMY IN THE BELLEVUE HOSPITAL MEDICAL COLLEGE, NEW YORK; FELLOW OF THE NEW YORK ACADEMY OF MEDICINE; MEMBER OF THE MEDICAL SOCIETY OF THE COUNTY OF NEW YORK; CORRESPONDENT OF THE ACADEMY OF NATURAL SCIENCES OF PHILADELPHIA, ETC., ETC.

NEW YORK:
D. APPLETON AND COMPANY,
549 & 551 BROADWAY.
1878.

COPYRIGHT BY
AUSTIN FLINT, Jr., M.D.,
1878.

PREFACE.

At the present time, there are few questions relating to physiology, of greater interest and importance than the one which is the subject of this essay. Since the publication of the experiments of Fick and Wislicenus, in 1866, a large number of observations have been made, which are brought forward as evidence that the muscular system of a fully-developed man or other animal is simply a perfected mechanical apparatus, like an artificially-constructed machine, which accomplishes work, not at the expense of its own substance, the material consumed being restored by food, but by using the food itself, the force-value of which can be accurately calculated, as we can calculate the dynamic value of the fuel consumed in a steam-engine. These observations have led some physiologists to adopt a kind of materialism, the fundamental idea of which is that the only matter concerned by its transformations in the production of muscular force is food. If a theory, with this idea as its basis, could be substantiated, it would indeed be an advance in positive knowledge, so great that its importance could hardly be over-estimated; and it is not surprising that the simplicity of the explanations of various physiological processes afforded by such an hypothesis should bring to its support many earnest and able advocates.

Since the enunciation of the laws of the correlation and conservation of forces, which are now almost universally accepted, it has seemed impossible to successfully controvert the notion that every manifestation of force in animal bodies, not excluding man, is dependent upon some kind of transformation of matter. Physiologists cannot comprehend the idea of the existence of any force unconnected with material changes, any more than it is possible to conceive of the absolute destruction of an atom of matter or of the generation or creation of something out of nothing.

Taking Nature as she now appears to us, there seems to be little or no basis for what may be termed an immaterial physiology. The researches which I have made into the question of the source of muscular power are not in any way opposed to the known relations between matter and force; they have been directed simply toward the solution of the problem whether the food be concerned directly, by its transformations, in the production of muscular power, or whether muscular effort involve changes in the muscular substance itself, this substance being destroyed as muscular tissue, discharged from the body in the form of excrementitious matter, and the waste being repaired by food. The gravity of this problem can be appreciated when it is remembered that complete and able treatises on physiology have lately been written upon the basis of the idea that food is directly concerned in the production of force, and that the muscular system, like the parts of a steam-engine, has no relation to the force developed, except that it consumes food and transforms it into energy, as a mechanical apparatus consumes fuel.

A logical method of inquiry to apply to this question is to disturb the natural balance between ordinary muscular work and the quantity of food, by increasing the work; then to calculate the income and outgo of matter and to ascertain, if possible, what is consumed in the production of force over and above that which can be accounted for by the food taken, assuming that this food is used either in repairing the muscular tissue consumed in the work or in the direct production of the work itself. If it can be shown by such a method of inquiry that excessive and prolonged muscular work consumes a certain amount of muscular tissue, it then becomes a question whether such work involve processes of destruction and nutrition of muscular substance, differing in kind as well as in degree from those which take place in ordinary muscular effort. But I shall not attempt here to prejudge any of the questions that will be involved in the discussion of the facts that I have at my command.

This essay appeared in the *Journal of Anatomy and Physiology*, Cambridge and London, for October, 1877, and, by some inadvertence, I did not receive the proofs before it was printed. The typographical errors, both in the figures and the text, were quite important. As it is, I have attempted, in this publication, to present an accurate statement of my own observations and what seem to me to be the logical conclusions to be drawn from these as well as from experiments made by others upon the human subject under conditions of rest and of muscular exercise.

NEW YORK, *December*, 1877.

THE SOURCE

OF

MUSCULAR POWER.

"It is now an established doctrine that force, like matter, can be neither created nor destroyed. The different forms of force are mutually convertible the one into the other, but they have their definite reciprocal equivalents, and in the transmutation the existing force undergoes no increase or decrease. The force liberated by a certain amount of chemical action will produce a certain amount of heat, and this, in its turn, may be made to accomplish a certain amount of mechanical work. The chemical action has its representative amount of heat, and the heat its representative amount of mechanical work; and the relative value of each is susceptible of being expressed in definite terms. It has been ascertained, for instance, that the force derived from chemical action which will raise the temperature of a pound of water 1° Fahr. will, under another mode of manifestation, lift 772 pounds one foot high; 772 foot-pounds are then said to constitute the dynamic equivalent of 1° of heat of Fahrenheit's scale.

"What is true of force in the inorganic world is equally applicable in the organic. The force manifested by living beings has its source by transmutation from other forms which have preëxisted. The food of animals contains force in a latent state. Properly regarded, food must be looked upon, not simply as so much ponderable matter, but as matter holding locked-up force. By the play of changes occurring in the body the force becomes liberated, and is manifested as muscular action, nervous action, assimilative, secretory, or nutritive action, heat, etc."

THE above is quoted from an article by Dr. F. W. Pavy, published in *The Lancet* for November 25, 1876. It contains a proposition which, if happily it were true, would mark an advance in our positive knowledge of animal mechanics, the importance of which could hardly be over-estimated—reducing our ideas of the physiology of muscular power to a degree of exactness and simplicity most attractive as well as desirable.

I do not propose to discuss here the law of the correlation and conservation of forces, as developed by researches in physics and inorganic chemistry; but it seems to me that the unreserved and absolute application of this law to the mechanics of the living body, particularly in their relations to the source of muscular power, is a question for careful physiological investigation, and not one that can be accepted simply upon the basis of analogy. It is easy to follow the various chemical actions induced in inorganic matters, to measure the heat produced and

GENERAL CONSIDERATIONS. 11

calculate the corresponding equivalents of force; but, while we can observe these processes accurately and without serious difficulty, when we come to study the various changes and transmutations which organic matters undergo in the animal body, we meet with problems which are, perhaps, the most intricate and complicated in Nature.

If the proposition advanced by Dr. Pavy were a legitimate and logical deduction from physiological investigations, nothing could be more simple than the mechanism of muscular power, and few would venture to contradict his views. This, however, does not seem to me to be the case. Dr. Pavy's proposition is apparently assumed to be true at the outset, as a condition precedent to his course of reasoning upon the results obtained by his observations. When the physiological data do not coincide with the theory under the influence of which his deductions seem to have been made, the error is assumed to be in the imperfection of the observations made by others as well as by himself. There is no suggestion that the theory itself may be faulty. Here seems to be the oft-repeated error of attempting to accommodate experimental facts to a law which is assumed to be invariable in its manifestations and exact in its applications; it is reasoning that a proposition, true as regards the inorganic kingdom, must be applicable absolutely to living bodies; it is bringing forward as evidence that the law is correct, arguments and deductions based mainly upon the assumption of the truth of the proposition involved.

In the study of animal physiology, we constantly meet with phenomena which are analogous to nothing with which we are acquainted in the inorganic world; and processes, which at first seem to be simple enough in their explanation, have been afterward ascertained to be of the most complex character. A notable instance of the latter is to be found in the history of the connection between respiration and the production of animal heat. In 1775, Lavoisier, who is justly regarded as one of the greatest chemists that ever lived, ascertained that the gas obtained by decomposing the oxide of mercury "was better fitted to maintain the respiration of animals than ordinary air."[1] Two years after, he confined animals under a bell-glass, and, after their death, showed that oxygen had been consumed and carbonic acid produced.[2] He afterward com-

---

[1] LAVOISIER, *Mémoire sur la nature du principe qui se combine avec les métaux pendent leur calcination, et qui en augment le poids. Hist. de l'Acad. Roy. des Sciences*, année 1775; Paris, 1778, pp. 521 and 525.

[2] *Expériences sur la respiration des animaux. Ibid.*, année 1777; Paris, 1780, p. 183.

pared the changes which take place in the air in respiration with the changes produced by the combustion of carbon and advanced the theory that heat is produced by a process in the animal body analogous to combustion. "Respiration is merely a slow combustion of carbon and hydrogen, which is in every way similar to that which takes place in a lighted lamp or candle; and, from this point of view, animals which respire are true combustible bodies which burn and consume themselves."[1] This was the origin of the theory which was afterward developed by Liebig into the doctrine of the hydrocarbons as calorific matters, and nitrogenized substances as plastic elements of food.

While the theories of Lavoisier and of his followers very greatly advanced our knowledge of the respiratory processes, the more modern researches of Bernard, Brown-Séquard, and others, with regard to the influence of the nervous system upon calorification, local variations in animal temperature, etc., showed that the production of animal heat is one of the phenomena incident to the general process of nutrition and that it is not due simply to oxidation of hydrocarbons. Although oxygen is consumed, carbonic acid is produced, and heat is

[1] Séguin et Lavoisier, *Premier mémoire sur la respiration des animaux. Hist. de l'Acad. Roy. des Sciences*, année 1789; Paris, 1793, pp. 570 and 571.

generated in the bodies of animals, we by no means understand all of the intermediate processes between the appropriation of oxygen by the tissues and the production of carbonic acid. We cannot raise the temperature of animals above the normal standard by increasing the supply of hydrocarbons in the food, nor can we arrest the production of heat by depriving animals of the so-called calorific principles. Physiologists have long since recognized the fact that the processes of combustion, such as we are familiar with in the inorganic world, are so far modified in the living body that the term "combustion," as applied to animal processes, means merely the appropriation of oxygen and not a simple chemical action resulting in the formation of carbonic acid and water and the production of a definite amount of heat.

Applying the lesson which should be learned from the progress of the theories of animal heat, to the study of the source of muscular power, physiologists, as it seems to me, should carefully study all of the facts known with regard to the development, nutrition, and disassimilation of muscular tissue. They should carefully weigh these facts before advancing a complete mechanical theory, in which food is regarded as "matter holding locked-up force," and calculating the heat-units and the

GENERAL CONSIDERATIONS. 15

foot-pounds of force necessarily contained in various alimentary principles.

No one can say why the growth of muscle and the development of muscular power is restricted within certain limits, no matter how much food may be taken and digested; no one can give a reason why a man becomes on an average five feet and eight inches high and weighs one hundred and forty pounds and then maintains about that standard throughout his adult life; why the limit of his muscular endurance is fixed; why, after exhausting one set of muscles, he is still capable of severe work with another; why the animal machine necessarily wears out and the being dies within a certain period; or why a man, with proper physical "training," becomes capable of greater and more prolonged muscular effort, with precisely the same food, than the same man out of training. These questions cannot be answered under the assumption that the animal mechanism is like a steam-engine, using food as fuel, which food produces a certain amount of work. If a man should take a certain amount of food and do no work except that required to maintain circulation, respiration, and assimilation, there is no evidence that the force "locked up" in the food is evolved in the form of heat; and the heat produced in the body is actually less than under exercise, as is well known. If it be assumed that the force "locked up" in food cannot be destroyed, what becomes of this force when the food undergoes its ordinary transmutations and there are no manifestations of force in work performed? Why does the muscular system need rest and recuperation after prolonged exertion, if it be simply the food which has its force liberated, and not the muscular system which wears itself to the point of exhaustion? When a steam-engine performs a certain amount of work, a part of the heat of the fuel is changed into mechanical power. When the same amount of fuel is consumed and no work is done by the machine, the heat is evolved and no part of it is transformed into mechanical power.

The only way in which certain of these questions can be answered is by making experiments upon the living organism under physiological conditions. The results of such experiments should be carefully studied, and then, and then only, may a comparison properly be made between a living mechanism and a machine artificially constructed and operated by methods with which we are perfectly well acquainted.

It would be illogical and unphilosophical to assume at the outset that the same methods operate in living bodies as in machines of our own construction, however attractive such an artificial simplicity of explanation might appear. We may thoroughly understand the principles of ordinary mechanics, but we should not necessarily assume that the results of physiological observations and experiments are faulty and imperfect, for the sole reason that they are not in accord with such principles as we comprehend from a study of the forces of inorganic Nature. We should rather humbly endeavor to discover the laws of living processes, and not seek to force such laws as we comprehend to apply absolutely to animal mechanism. Our physiological facts, if definite and well established, should be treated as facts not to be distorted into arguments in favor of laws which we have enacted rather than discovered. Whatever facts we possess in physiology have been ascertained by long, patient, and difficult research, and have generally been established after much discussion and controversy and the apparent opposition of conflicting observations.

The various functions of the human body are so dependent upon and so closely related to each other, that it is difficult, in the present condition of science, for any one but a physiologist to accurately and justly weigh the evidence bearing upon important physiological questions. In our literature, one can often find what appears to be good authority for diametrically opposite views, between which an intelligent judgment can be formed only with a profound knowledge of general physiology and with a simple and earnest desire to ascertain the truth as truth and not necessarily in its relations to the laws of physics and mechanics. The laws of inorganic Nature are usually definite and undisputed. Physiologists may use such laws and the facts which correspond to them in their applications to the animal functions; but observations with regard to the processes in animated Nature are not so fixed and definite. Physicists, chemists, and mechanicians often work in the field of physiology; but they generally remain physicists, chemists, or mechanicians, and they often take from physiology upon authority that only which suits their theories of what physiology should be, opposite observations to the contrary notwithstanding.

The question of the source of muscular power is not necessarily one of chemistry, physics, or mechanics. We must approach this question physiologically. Reasoning first upon what we know of the development, nutrition, and disassimilation of muscular tissue, we should endeavor to ascertain what are the phenomena which attend the exercise of muscular power. If we can arrive at any definite experimental facts by this method, we may then consider the heat-value and the force-value of food. No positive and stable doctrine can be established by assuming that the equivalents of heat and force ascertained by treating food as inorganic matter are to be rigorously applied to those mysterious changes which food and tissue undergo in the living organism. With such a method, there would be no necessity for what may be regarded as purely physiological observations, and the science of physiology would be reduced to a kind of materialism, in which the development of force would involve changes in matter from without instead of the matter of the organism itself.

Treating, as I shall endeavor to do, the question under consideration from a strictly physiological point of view, I shall first draw attention to what we know of the nutrition and development of muscular tissue, and then discuss certain observations upon the human subject during repose and during the exercise of muscular power, in which the physiological conditions appear to have been fulfilled.

Nutrition and Development of the Muscular System.

I do not propose, in this connection, to consider the original development of the muscular tissue, but shall endeavor to show how the nutrition of muscles may be promoted by diet and exercise so as to develop them to the maximum of size, strength, and power of endurance. These ends are accomplished by what is known as the process of "training."

The muscles of a person who takes little or no exercise are usually small, soft, and of comparatively little power. The endurance, or capacity for prolonged muscular effort, is not great. The "wind" is deficient; that is, any unusual muscular effort is likely to produce temporary distress in breathing, and exhaustion. In healthy persons who habitually take but little exercise, coincident with a want of development of the muscular system, there is generally more or less fat in those situations in which fat is usually deposited. The difficulty in breathing after unusual exertion is in part due to the deposition of fat in the omentum, which interferes with the free action of the diaphragm, in part to a want of habit of vigorous action of the respiratory muscles generally, and to what we may rather indefinitely term nervous exhaustion. In the exercise of running or fast walking, a person in this condition usually carries a certain weight of inert matter in the form of fat, which increases the demands upon his muscular system. By a judicious process of training, however, the fat may be reduced to the minimum, the muscles, or working part, may be developed to the maximum, the general tone of the nervous system is improved, and the habit of efficient respiration is acquired, so that this process is carried on easily and in a way to supply the increased quantity of oxygen required during prolonged muscular effort.

When any particular set of muscles is persistently and systematically exercised, these special muscles are developed out of proportion to the rest of the body, and they become capable of unusual power and effort. The general effects of such systematic exercise are the following: An increased consumption of oxygen, an increased elimination of carbonic acid, improved digestion, an increased demand for food, and a diminished tendency to the accumulation of fat, while the muscles exercised become harder and are sometimes increased in volume. Sometimes, when the muscles are originally large and soft, they are diminished in volume by exercise. During exercise, there is an increased production of heat in the body, but the animal temperature is maintained at the normal standard by increased evaporation from the general surface. During exercise, also, the general circulation becomes more active. The influence of exercise upon the elimination of excrementitious matters, particularly urea, is the main question for discussion and will be fully considered farther on.

The effects of exercise upon particular muscles are the following: A local elevation of temperature, an increased supply of blood, and a condition which is followed by increased activity of nutrition with a diminution in the quantity of interstitial fat.

To bring the muscular system to the maximum of power and endurance, the exercise must be carefully directed to that end, and the diet must be judiciously regulated. In the first place, the exertion should rarely be so severe or prolonged as to induce anything more than a temporary exhaustion, followed promptly by an agreeable reaction and a sense of fatigue readily relieved by repose. In training for a feat of strength involving a short and supreme effort, the daily exercise may culminate in an effort a little less severe than that which it is desired finally to attain. For feats of endurance, the daily exercise should be less severe but more prolonged. A complete rest for twenty-four or forty-eight hours is desirable just before the feat to be accomplished is attempted, and the attempt should never be made while the digestive process is in full operation.

It is surprising how short a period of vigorous exercise daily will develop an approach to the maxi-

DEVELOPMENT OF THE MUSCULAR SYSTEM. 23

mum of muscular power. At the age of forty years and weighing one hundred and eighty-three and three-quarter pounds without clothing, I myself accomplished the feat of raising with one hand above my head and standing erect with the arm straight under a dumb-bell weighing one hundred and eighty and one-half pounds. This was done by exercising about half an hour daily for six days in the week, paying no special attention to the diet. The course of training for this special feat of strength was continued for about five months. At the beginning of the five months, I could easily "put up" a dumb-bell weighing one hundred and sixty-five pounds. I never trained specially for any feat of endurance, when attention to the diet would probably have become necessary. I believe that one hour a day of vigorous exercise, with proper attention to diet, will efficiently train a well-formed and healthy man for any reasonable feat of strength or endurance.

The training diet recognized everywhere upon empirical and scientific grounds consists of the most nutritious meats, in such a form as to be easily digested, eggs, liquids in small quantity, and little or no fat, sugar, starchy matters, alcohol, tea, coffee, tobacco, or articles which appear to retard disassimilation. The quantity of food of the kind indicated is not restricted except within the limits of good digestion. Nervous excitement of all kinds is avoided, the functions of the skin are promoted as much as possible, and natural and sufficient sleep is indispensable. A man in proper training is supposed to live for a time a purely physical life, with no end in view except the perfect development of his muscular system. He should experience a sense of high physical enjoyment in his course of training. How different is it with the poor, who are hard worked and insufficiently fed! The exertion in such instances is depressing and exhausting; the muscular system becomes enfeebled; the deficiency in nourishment frequently induces an unnatural craving for alcohol and other stimulants; the system loses its power to resist disease; the work, instead of developing power of endurance, reduces the general tone of the system, and existence becomes almost a burden. On the other hand, a laboring-man, moderately worked and well fed, is frequently the very type of animal vigor.

It may now be inquired how the physiological exercise of the muscular system, with sufficient alimentation, affects the nutrition of the muscles themselves. Exercise undoubtedly increases, within certain limits, the power of the muscular system to assimilate the matters required for its proper nu-

DEVELOPMENT OF THE MUSCULAR SYSTEM. 25

trition and the full development of strength and endurance; but the exercise must be periodic and followed by sufficient intervals of repose and recuperation. These periods of repose and the assimilation of a proper amount of nutriment are recognized as absolutely necessary.

The question as to whether exercise actually consumes the muscular substance is the main one for consideration, the doctrine of some physiologists being that the consumption of matter in muscular work involves the elements of food and not the substance of the muscular tissue itself. I shall discuss this question rather briefly in this connection, leaving it to be answered more fully hereafter by experimental data. Take, as an illustration, the case of a man training for an athletic contest, such a course as I have often observed and in one instance directed. The greatest part of the actual muscular work takes place within two or three hours of the twenty-four; about eight hours are devoted to sleep, and the rest of the day is occupied in recreation, eating, etc. The fat of the body becomes so far reduced that it finally almost disappears from the omentum, the subcutaneous tissue, and the interstices of the muscles. No sugar and but little starch and fat are taken, the diet being almost exclusively nitrogenized. No albuminoid matters are discharged as such from the body, and the chief principle eliminated that contains nitrogen is urea. Taking the time actually employed in muscular work as about two hours, there is then a decided increase in the process of disassimilation, as far as this is to be measured by the elimination of carbonic acid, and the consumption of oxygen is proportionately increased; but, if nitrogenized food be not taken in sufficient quantity, there is a sense of hunger, and the capacity for muscular work is diminished. The sense of hunger has its seat, not in the blood, but rather in the system at large, where the nervous system can be brought into action. All the food taken is probably digested and absorbed in six or eight hours, and the elimination of nitrogen by the kidneys is going on constantly.

It is almost impossible to imagine that the nitrogen eliminated in the urine during the long periods of the day when there is complete rest of the general muscular system represents only the work of the heart (which weighs but eight or ten ounces) and the action of the muscles in tranquil respiration, and that there is no disassimilation of the muscular tissue generally; and there is absolutely no evidence that the nitrogenized elements of food are taken into the blood and there directly changed

DEVELOPMENT OF THE MUSCULAR SYSTEM. 27

into urea. It is stated by Sappey that the muscular system equals about two-fifths of the weight of the entire body in a well-proportioned man,[1] and in a man in high muscular training the proportion must be much larger. According to this estimate, the muscles of a man weighing one hundred and forty pounds would weigh about fifty-six pounds, and about eighteen pounds of blood circulate through these muscles constantly, the whole mass of blood passing through the heart about once a minute. With this constant passage of blood through the muscular tissue, it is probable that something more of an interchange takes place between the blood and the muscular substance than the mere consumption of oxygen and giving off of carbonic acid. It is true that the muscular tissue contains little or no urea; but the liver during life contains little or no sugar, while sugar is being constantly formed in the liver, is as constantly washed out by the current of blood, and always exists in the blood of the hepatic vein.

The large amount of nitrogenized food taken during training is necessary in order to maintain the muscular system at a certain standard of weight. If this be the fact, the muscles must be constantly losing substance by disassimilation and as constantly

[1] SAPPEY, *Traité d'anatomie*, Paris, 1868, tome ii., p. 6.

repairing themselves by matters appropriated from the blood. As the muscles constitute the largest part of the nitrogenized constituents of the body, it is reasonable to suppose that the excretion of urea, which is the chief nitrogenized excrementitious substance, represents, more or less fully, the activity of muscular disassimilation. In support of these ideas is the following fact, which has been long recognized:

For sake of illustration I suppose that a pugilist, weighing two hundred pounds, agrees to fight in three or six months at a weight of one hundred and sixty-five pounds or under, at which latter weight, he judges from experience that he can maintain his strength and endurance. He puts himself in "training" and first eliminates all of his fat by exercise, sweating, and a nitrogenized diet, which may easily be done. When this has been accomplished, although he has no fat, he weighs one hundred and seventy-five pounds, or is ten pounds "over-weight." The object now is to reduce the weight to the prescribed standard, and in this process to weaken the muscular force and endurance as little as possible; but it is evident that this cannot be done without reducing the weight of the muscles. To accomplish this, the usual course is to exercise violently and to promote profuse sweating, at the same time restrict-

DEVELOPMENT OF THE MUSCULAR SYSTEM. 29

ing the liquids as much as possible. In this way, as the loss by perspiration is not supplied, the weight of the body must diminish. The deficient quantity of water seems to prevent the full supply of reparative matter to the muscles, while it does not interfere so much with their disassimilation. At all events, it is evident that the loss of weight involves the muscles almost exclusively, and this is probably due primarily to the excessive exercise. With this violent exercise, the muscular system might be reduced in weight by restricting the quantity of nitrogenized food; but experience has shown that this course involves much greater loss of strength than the reduction of weight by restricting the quantity of liquids.

Practically, it has been found to be very difficult to keep the weight of the body much below the normal standard for any considerable length of time; and an increase in the quantity of liquids taken will add several pounds to the weight in the course of a few hours. Such reduction of weight, however, has its limit; and it has very often occurred that men have miscalculated the effects of this severe course of training, and, although they have gone into the ring at the proper weight, and apparently in very " fine condition," they have been utterly incapable of making a contest.

Some persons, whose muscles are small, and who have no tendency to the accumulation of fat, increase very considerably in weight under the ordinary course of training.

A steam-engine is not "trained" to accomplish a certain amount of work. A machine of this kind is perfected in all of its parts, and is so constructed as to be of sufficient strength to overcome such resistance as it is likely to meet. It is simply an apparatus for transforming heat furnished by fuel into useful force, and it is nothing without fuel. Man, on the other hand, is a living being, developed from a fecundated ovum of microscopic size, by a process which we have hardly begun to comprehend. In his growth, the various tissues and organs have the power of appropriating materials for development, when they are presented in an appropriate form and under proper conditions. There is no reason to suppose that the nature of this process of nutrition radically changes when the being reaches adult life, and there is no single reasonable argument in favor of such a view. A man may take a certain quantity and kind of food, and still, without training, be able to perform only a certain amount of work. After proper training, with precisely the same food, he can develop greatly-increased power. His span of life is definitely fixed, and no amount of care can prevent those retrograde organic changes which result in death.

Assuming that a proper system of training is essential to perfect development of the machinery of the muscular system, the simple question is whether, in the perfected muscular system of the adult, force be generated by changes of the muscular substance or whether the force be due to the direct transformation of elements of food. In other words, is the muscular substance an apparatus for transforming the force locked up in food into power, or are the muscles themselves consumed, the elements of food being used for their repair? These questions may be resolved by little more than a single experimental line of inquiry: Does physiological exercise of the muscular system increase the elimination of nitrogenized excrementitious principles?

Relations of the Muscular System to the Elimination of Nitrogen.

There seems, at the first blush, to be good ground for supposing that the elimination of nitrogen is closely related to the physiological wear of muscular tissue, for several reasons. The muscular system may be, under certain circumstances, the only part of the body that is materially affected by exercise. In a man of ordinary development, the muscular system constitutes at least two-fifths of the total weight.[1] Fat may disappear almost entirely from the body, and the food may be restricted to nitrogenized matters, without disturbing nutrition. These matters are never discharged from the body as albuminoids, but the nitrogen is eliminated mainly in the urea. Under such conditions, and with a varying amount of exercise, it is the muscular system only which presents any considerable changes in weight.

Supposing the fat of the body to be reduced to its minimum proportion—and it usually constitutes but about one-twentieth of the total weight[2]—there are no other parts that can be affected by exercise; for there is no reason to suppose that the nervous system, the abdominal, thoracic, or pelvic viscera, the skin, bones, or tendons, present any immediate changes in weight as the result of muscular exertion. Take the case of Weston, the pedestrian, who was under my observation in 1870, and who weighed a little more than one hundred and nineteen pounds just before he began a walk of five consecutive days. He must have had at least forty-eight pounds of muscular tissue, and he was reduced in weight, dur-

---

[1] SAPPEY, *Traité d'anatomie*, Paris, 1868, tome ii., p. 6.
[2] CARPENTER, *Principles of Human Physiology*, Philadelphia, 1876, p. 66.

ing a walk of two hundred and seventy-seven miles in four consecutive days, to one hundred and fourteen pounds.[1] During this time, he probably consumed five pounds of muscular tissue which could not be repaired by food, or about ten per cent. of the total weight of muscle. It might be assumed that he consumed this amount of muscular substance because the food taken was insufficient to produce the force exerted; but it is more reasonable to suppose that he lost muscular weight because the food could not repair the excessive waste engendered by the extraordinary amount of work accomplished. However this may be, our views must rest upon the experimental answer to the question whether or not muscular exercise increase the elimination of nitrogen from the body, irrespective of theoretical considerations relative to the force-value of food, which are derived solely from physical and chemical observations and which do not take into account the complex and imperfectly-understood processes of nutrition, the mysterious influences of the nervous system, or the intricate series of phenomena which are intermediate between the appropriation of oxygen and the production of carbonic acid by the tissues.

In discussing the various experiments that have

[1] This was the greatest loss of weight observed at any time during the walk of five days.

been made with regard to the influence of muscular exercise upon the elimination of nitrogen, I shall confine myself to those which refer to the human subject. It is evident that such observations must be made when the subjects of the experiments are under strictly physiological conditions, especially as regards alimentation and general nutrition. It is evident, also, that the direct influence of food upon the excretion of nitrogen must not be neglected. To meet this latter requirement, it would seem proper, in estimating the amount of nitrogen discharged under various conditions, to calculate the proportion of nitrogen discharged to the nitrogen of food. This, however, has not been done in all of the experiments which I shall discuss.

Experiments of Liebig, Lehmann, Fick and Wislicenus, and Parkes.

The experiments of Fick and Wislicenus, which were published in 1866, are supposed by some to have revolutionized the ideas of physiologists with regard to the significance of the excretion of nitrogen. Before that time, the idea of Liebig, which is expressed in the following quotations, was generally adopted:

"Boiled and roasted flesh is converted at once into blood; while the uric acid and urea are derived from the metamorphosed tissues. The quantity of these products increases with the rapidity of the transformation in a given time, but bears no proportion to the amount of food taken in the same period. In a starving man, who is in any way compelled to undergo severe and continued exertion, more urea is secreted than in the most highly-fed individual if in a state of rest."

The last statement contained in the above quotation is very broad, and it does not appear to be made upon the basis of direct experiment.

Again, Liebig makes the general statement that "the amount of tissue metamorphosis in a given time may be measured by the nitrogen in the urine."[1]

The doctrine thus enunciated by Liebig was modified, a few years later, by the researches of Lehmann, who showed, by observations upon his own person, that, other conditions being equal, the character and quantity of food modified very greatly the elimination of urea, as is seen by the following quotation:

"My experiments show that the amount of urea which is excreted is extremely dependent on the nature of the food which has been previously taken. On a purely animal diet, or on food very rich in nitrogen, there were often two-fifths more urea excreted than on a mixed diet; while, on a mixed diet, there was almost one-third more than on a purely vegetable diet; while, finally, on a non-nitrogenous diet, the

[1] LIEBIG, *Animal Chemistry*, London, 1843, pp. 138 and 245.

amount of urea was less than half the quantity excreted during an ordinary mixed diet."

Lehmann farther states, however, that, upon a uniform diet, the elimination of urea is increased by muscular exercise.[1]

In 1866, Fick and Wislicenus published their account of experiments made in ascending one of the Alpine peaks, the Faulhorn, about 6,500 feet high. These experiments were undertaken with the view of showing that severe and prolonged muscular effort could be accomplished upon a non-nitrogeneous diet. The two experimenters took no albuminoid food from mid-day on August 29th until seven P. M. on August 30th. The experiments proper began on the evening of the 29th, at a quarter-past six P. M., by a complete evacuation of the bladder. The urine from this time until ten minutes past five on the morning of the 30th (about eleven hours) was collected, and called the "night urine." The ascent began at ten minutes past five and occupied eight hours and ten minutes. The urine passed during this period was collected as "work urine." The urine for five hours and forty minutes after the ascent was collected as "after-work urine." The urine from seven P. M., August

---

[1] LEHMANN, *Physiological Chemistry*, Philadelphia, 1855, vol. i., p. 150.

30th, until half-past five A. M., August 31st, was collected and designated as "night urine." The results of the examinations of these specimens in the two persons were nearly identical. The following is the estimate of the elimination of nitrogen per hour during the different periods:

	Fick.	Wislicenus.
During the night, 29th to 30th,	0·63 gramme.	0·61 gramme.
During the time of work,	0·41 "	0·39 "
During rest after work,	0·40 "	0·40 "
During the night, 30th to 31st,	0·45 "	0·51 "

From these results, Fick and Wislicenus conclude that muscular exercise does not necessarily increase the elimination of nitrogen; that the substance of the muscle itself is consumed in insignificant quantity; and that the muscular system is a machine, consuming, in its work, not its own substance, but fuel, which is supplied by the food. The most efficient fuel Fick and Wislicenus consider to be non-nitrogenized food; the results of its consumption being force (or work), heat, and carbonic acid. They adopt the view "that the substances, by the burning of which force is generated in the muscles, are not the albuminous constituents of the tissues, but non-nitrogenous substances, either as fats or hydrates of carbon."

"We might express this doctrine by the following simile:

A bundle of muscle-fibres is a kind of machine consisting of albuminous material, just as a steam-engine is made of steel, iron, brass, etc. Now, as in the steam-engine coal is burnt in order to produce force, so, in the muscular machine, fats or hydrates of carbon are burnt for the same purpose. And in the same manner as the constructive material of the steam-engine (iron, etc.) is worn away and oxidized, the constructive material of the muscle is worn away, and this wearing away is the source of the nitrogenous constituents of the urine. This theory explains why, during muscular exertion, the excretion of the nitrogenous constituents of the urine is little or not at all increased, while that of the carbonic acid is enormously augmented; for, in a steam-engine, moderately fired and ready for use, the oxidation of iron, etc., would go on tolerably equably, and would not be much increased by the more rapid firing necessary for working, but much more coal would be burnt when it was at work than when it was standing idle." [1]

The question under consideration is not materially advanced or modified by the experiments of Frankland [2] or of Haughton,[3] who adopt fully the views of Fick and Wislicenus.

In 1867, experiments were made by the late Dr. Parkes upon two soldiers, with the view of control-

[1] FICK AND WISLICENUS, *On the Origin of Muscular Power.—London, Edinburgh, and Dublin Phil. Mag.*, London, January to June, 1866, vol. xxxi. pp. 492 to 501.

[2] FRANKLAND, *On the Origin of Muscular Power.—London, Edinburgh, and Dublin Phil. Mag.*, London, July to December, 1866, vol. xxxii., p. 182, *et seq.*

[3] HAUGHTON, *The Lancet*, London, August 15, 22, and 29, 1868.

ling the experiments of Fick and Wislicenus by observations upon a more extended scale.[1] These experiments were continued for a period of eighteen days, and they certainly seem to show an increase in the elimination of urea, attributable to muscular exercise. The extraordinary exercise taken was a walk of 23·7 miles on one day, and 32·78 miles on the day following. During these two days, upon an exclusively non-nitrogenized diet, the elimination of nitrogen was slightly increased over a period of two days of rest with a non-nitrogenized diet. In an analysis of a recent course of lectures delivered by Dr. Parkes at the College of Physicians, London, it appears that he is disposed to take a view of the subject between the two extremes; viz., that the muscular system is able to accomplish work by the consumption of non-nitrogenous food; that exercise does, however, slightly increase the elimination of urea, and that, during exercise, a small portion of the muscular substance is consumed; but he holds that the variations in the quantity of nitrogen eliminated are almost entirely dependent upon the amount of nitrogen contained in the food.[2]

[1] PARKES, *On the Elimination of Nitrogen by the Kidneys and Intestines, during Rest and Exercise, on a Diet without Nitrogen.*—*Proceedings of the Royal Society*, London, 1867, vol. xv., No. 89, p. 339, *et seq.*

[2] *Medical Times and Gazette*, London, March 15, 1871, p. 348.

In 1870, Liebig published an article in which he again discussed the question from his own point of view. He analyzed very fully the experiments of Parkes and found in the results fresh testimony in favor of his view that the increase in the elimination of nitrogen as a consequence of muscular exercise is not limited to the period of exertion but continues for some time after.[1] On the other hand, Voit published, also in 1870, an elaborate paper, reviewing the publications on this question that had appeared for the past twenty-five years.[2] Neither of these papers, however, has added to the sum of physiological knowledge by the contribution of new experimental facts; but they are interesting as expressing the arguments upon two opposite sides, and they illustrate the necessity of new observations, in which some of the important omissions in the experiments hitherto made may be supplied.

A fatal objection, in my opinion, to the observations of Fick and Wislicenus, which constitute the starting-point of the new theory of the origin of muscular power, is the fact that the experiments were made when the system was not under physiological conditions.

[1] LIEBIG, *The Source of Muscular Power.—Pharmaceutical Journal and Transactions*, London, 1870, third series, part ii., p. 161, and part iii., pp. 181, 201 and 222.

[2] *Zeitschrift für Biologie*, München, 1870, Bd. vi., S. 305, *et seq.*

For a period of thirty-one hours, these two experimenters took no albuminoid food; and, within that time, they occupied eight hours and ten minutes in making an ascent of 6,500 feet. It cannot be assumed that they were in a proper condition, as regards alimentation, to perform this work; and it is illogical to conclude, as the result of such observations, "that the substances, by the burning of which force is generated in the muscles, are not the albuminous constituents of the tissues, but non-nitrogenous substances, either as fats or hydrates of carbon." No attempt was made to measure the carbonic acid eliminated, or the part of the assumed changes in the non-nitrogenous matters consumed which was concerned in the production of heat, and no account was taken of the variations in the weight of the body. Again, it is well known that the change of the normal diet to a regimen of non-nitrogenous matters alone of itself diminishes very largely the excretion of nitrogen.

In addition to what I have already quoted from Lehmann, he states that "there is as much urea in the urine after a prolonged absence from all food (after a rigid fast of twenty-four hours) as after the use of perfectly non-nitrogenous food." Taking the results of the experiments upon Fick (and those upon Wislicenus are almost identical), for about eleven hours just preceding the ascent, the elimination of nitrogen per hour amounted to 0·63 of a gramme. During this time, and for about six hours before, no nitrogenous food was taken. During the succeeding eight hours and ten minutes, which were occupied in the ascent, the hourly elimination of nitrogen was 0·41 of a gramme, a reduction of about one-third. During five hours and forty minutes after the work, still without nitrogenous food, there was a farther, but a very slight, reduction in the nitrogen eliminated, the hourly quantity being 0·40 of a gramme. During the succeeding period of ten hours and thirty minutes, with a return to nitrogenous food, there was an increase in the nitrogen eliminated, amounting to a little more than twelve per cent. above the lowest point, the hourly quantity being 0·45 of a gramme. Lehmann has shown that, irrespective of muscular exercise, the elimination of nitrogen is diminished more than one-half by a non-nitrogenous diet. This would fully account for the diminution observed by Fick and Wislicenus.

Viewed physiologically, without reference to any particular theory of the origin of muscular power, the experiments of Fick and Wislicenus seem to have no great value. The only way in which they could have had any bearing upon the question under consideration would have been by comparing them with observations under the same conditions of diet, but with no muscular work. If Fick and Wislicenus had shown that the excretion of nitrogen, without muscular exercise, was diminished to a certain degree by non-nitrogenous diet, and that, with work, under precisely the same diet, the elimination was diminished to a greater degree than it had been without work, this would have shown that " muscular exercise does not necessarily increase the elimination of nitrogen." Without such observations upon the influence of a non-nitrogenous diet without work, I fail to understand how physiologists can accept the doctrine enunciated by Fick and Wislicenus, except upon the assumption of the correctness of a theory which rests entirely upon the basis of their experiments. Such methods of reasoning would certainly tend to retard our advance in positive knowledge.

Dr. Pavy, in his observations upon this subject, supplies the omission made by Fick and Wislicenus by experimental facts which really overthrow their theory.[1] Dr. Pavy gives an illustration where a person under ordinary conditions as regards exercise passed from a mixed to a purely non-nitrogenous

[1] *The Lancet*, London, December 23, 1876, p. 888.

diet. "During seventeen hours of May 5th, while upon mixed food, the urea eliminated represented 197 grains of nitrogen." (This equals 11·59 grains per hour.) "During twenty-three hours of May 6th, when only non-nitrogenous food, consisting of arrow-root, sugar, and butter, was taken, the nitrogen in the urea voided amounted to 187 grains" (this equals 8·13 grains per hour); "and for the twenty-four hours of the following day, the diet still consisting of non-nitrogenous food, it amounted to 136 grains." (This equals 5·66 grains per hour.) Under uniform conditions of exercise, then, non-nitrogenous food diminished the nitrogen excreted for the first twenty-three hours about one-third below the standard with an ordinary mixed diet. The non-nitrogenous diet continued through the succeeding twenty-four hours diminished the nitrogen excreted to about one-half of the normal standard. These results nearly coincide with those obtained by Lehmann.

The observations upon Fick during about eleven hours of a non-nitrogenized diet without work gave 0·63 of a gramme of nitrogen per hour. This period may be taken as corresponding with the first period of twenty-three hours in Dr. Pavy's observations, in which the hourly elimination of nitrogen was 8·13 grains, as the nitrogen in Fick's observation was undoubtedly reduced below the standard under a mixed diet. During the eight hours and ten minutes of work in Fick's observations, still with a non-nitrogenous diet, the nitrogen was reduced from 0·63 of a gramme to 0·41 of a gramme per hour, or about one-third. During the twenty-four hours of Dr. Pavy's observation, still with a non-nitrogenous diet, but with no variation in exercise, the nitrogen excreted hourly was reduced from 8·13 grains to 5·66 grains, or about one-third, nearly the same as in the observations upon Fick. At the end of the time calculated by Fick, the non-nitrogenous food had been taken for twenty-five hours and twenty minutes; viz., from noon on August 29th to twenty minutes past one on August 30th. At the end of the time noted for Dr. Pavy, the non-nitrogenous food had been taken for forty-seven hours, with no variation in the exercise. These observations, therefore, seem to show that the diminished excretion of nitrogen under a purely non-nitrogenous diet, which is certainly not a physiological condition, depends entirely upon the food.

The observations of Parkes, which were made " on a diet without nitrogen," are open to precisely the same objections and invite the same criticism as that which I have made upon the experiments of Fick and Wislicenus.

Experiments of Dr. Pavy.

The experiments of Dr. Pavy upon Weston, the pedestrian, in 1876,[1] were made upon the same person, by the same method, and under the same conditions as those which I made in 1870. The idea of my own experiments, which was adopted by Dr. Pavy in his observations, was to measure the proportion of nitrogen eliminated to the nitrogen of food, during rest and during extraordinary muscular exertion, with the view of ascertaining the influence of exercise upon nutrition and disassimilation. Dr. Pavy corrected some errors in method and calculation in my experiments, which I willingly accept and shall refer to farther on. I reserve, however, the discussion of my own experiments for the last, because it is from them that I shall make the deductions which will conclude this essay. Still, I shall compare Dr. Pavy's observations with mine in the course of the discussion.

The subjects of Dr. Pavy's experiments were the following:

1. A walk by Perkins, the pedestrian, of sixty-five and one-half miles in twenty-four hours.

---

[1] *The Lancet*, London, February 26, March 4, 11, 18, 25, November 25, December 9, 16, 23, 1876, and January 13, 1877.

2. A period of twenty-four hours of rest following the walk, after an interval of several days.

3. Two periods of twenty-four hours each, in which Weston, the pedestrian, walked, during the first twenty-four hours, one hundred miles, and during the second twenty-four hours, eighty and one-half miles.

4. A period of twenty-four hours of rest immediately preceding a walk by Weston of seventy-five hours.

5. Three periods of twenty-four hours each, in which Weston walked, during the first period, one hundred and four miles, during the second period, seventy-six miles, and during the third period, eighty-three miles.

6. A period of twenty-four hours of rest immediately succeeding Weston's walk of seventy-five hours.

7. Six periods of twenty-four hours each, immediately preceding a walk by Weston of six days.

8. Six periods of twenty-four hours each, in which Weston walked, during the first period, ninety-six miles; during the second period, seventy-seven miles; during the third period, seventy and one-half miles; during the fourth period, seventy-six and one-half miles; during the fifth period, sixty-seven miles; and during the sixth period,

sixty-three miles, making a walk of four hundred and fifty miles in six consecutive days.

9. Six periods of rest of twenty-four hours each, immediately succeeding Weston's walk of six days.

In these observations, the number of miles walked is given for each period, except for the periods of rest, when it is assumed that little or no exercise was taken.[1]

The body weight is given, except for the rest-periods before and after Weston's seventy-five hours' walk, the six periods of twenty-four hours each, immediately succeeding Weston's six days' walk, and Perkins's period of rest. For the six days preceding Weston's six days' walk, the weights are given as "without clothing," and they cannot be compared with the weights for the six days' walk, which are given as "in costume."

The nitrogen of food is given for all the periods except for the six days following Weston's six days' walk. It is not stated precisely how the estimates of the quantities of the different articles of food were made, but, as these are so important, I must

---

[1] For the six days' period of rest previous to the six days' walk, the following note is made by Dr. Pavy: "During the first two days Weston kept entirely to the house, and during the last four the amount of walking did not reach twenty miles. The results, therefore, will represent a state of comparative rest."—*The Lancet*, March 11, 1876, p. 392.

EXPERIMENTS OF DR. PAVY. 49

assume that they are approximatively correct. At least, it is fair to suppose that the errors in these estimates, if any errors exist, are uniform for all the different periods, and therefore that they do not affect the comparative results.

In my own observations, each separate article of food was placed upon a different plate, which was weighed before and after eating, the difference in weight giving the actual quantity consumed. This, of course, would eliminate bone and matter not actually eaten. I am led to call attention to this, for the reason that the average amount of nitrogen consumed daily by both Perkins and Weston for twelve days of walking was 556·79 grains. The average daily consumption of nitrogen by Weston, observed by me, for five days of walking, was 234·76 grains, less than one-half that noted by Dr. Pavy; and the daily average for ten days of rest was 390·18 grains. Dr. Pavy, however, notes that the food taken was much greater in quantity in his experiments.[1] In all of my experiments, Wes-

[1] After I had seen the first detailed account of Dr. Pavy's observations, which was published in *The Lancet* several months before the publication of the tabulated results and the conclusions, I made a rough estimate of the nitrogen of the food, using the proportions which I had taken for my observations upon Weston in 1870. According to this estimate, the average amount of nitrogen consumed daily by both Perkins and Weston for twelve days of walking was 460·05 grains. Dr. Pavy's estimate, published later,

ton took what he wanted, both as regards quantity and kind of food. It seems to me to have been an important omission on the part of Dr. Pavy that he did not note the weights of the body and the amount of nitrogen taken during the six days following the six days' walk, when the system was recuperating after the great exertion which had been made. The nitrogen of food was estimated

was very much higher, the daily average for the same periods being 556·79 grains. After Dr. Pavy had published his tabulated results and conclusions, I made a calculation of the nitrogen of food for the second twenty-four hours of the six days' walk, which gave, according to Dr. Pavy's estimate, the greatest quantity of nitrogen; viz., 826·43 grains. I endeavored to estimate the nitrogen according to Dr. Pavy's method, but I could not make more than 781·73 grains for the twenty-four hours. This is the only day for which I have endeavored to verify Dr. Pavy's calculations, and I cannot see why our results should differ so much. The quantities of food for this day are given in *The Lancet*, March 18, 1876, page 430, and the estimates of the proportion of nitrogen in the various articles of food are in *The Lancet*, November 26, 1876, page 742. The following is my own estimate, according to Dr. Pavy's figures, for that day.

SECOND TWENTY-FOUR HOURS OF WESTON'S SIX DAYS' WALK.

Quantities of Food.	Nitrogen of Food.
Cooked meat, 1 lb. 6¼ oz. (15·31 grs. of nitrogen per oz.),	344·475 grains.
Three yelks of eggs (6·45 grs. of nitrogen in each),	19·350 "
Four poached eggs (16·62 grs. of nitrogen in each),	66·480 "
Jelly, 1 pt. 3 oz. (2·62 grs. of nitrogen per fluid oz.),	49·780 "
Beef-tea from fresh meat, ¼ pt. (1·55 gr. of nitrogen per fluid oz.),	12·400 "
Liebig's extract, 4¼ oz. by weight, taken as beef-tea (37·73 grs. of nitrogen per oz.),	160·350 "

EXPERIMENTS OF DR. PAVY. 51

by Dr. Pavy by essentially the same methods as those employed by me, using mainly the calculations of Payen.

To represent the nitrogen excreted, Dr Pavy estimated the nitrogen contained in the urea and uric acid for each period in all of his observations, while I estimated the nitrogen contained in the urea and fæces. The excretion of uric acid was found to be much greater in Dr. Pavy's observations than in mine. This difference, although considerable, is not so great as to affect the conclusions in either case. Dr. Pavy objects to my

Quantities of Food.	Nitrogen of Food.
Brand's essence of beef, 1¼ oz. by weight (4·37 grs. of nitrogen per oz.),	5·462 grains.
Milk, 1¼ pt. (2·22 grs. of nitrogen per oz.),	47·360 "
Oatmeal, 3½ oz. in the form of gruel (8·53 grs. of nitrogen per oz.),	29·855 "
Bread, 3 oz. in the form of dry toast (4·68 grs. of nitrogen per oz.),	14·040 "
Bread, spread with butter, 2½ oz. (1·40 gr. of nitrogen allowed for butter),	13·100 "
Potatoes, 3 oz. (1·44 gr. of nitrogen per oz.),	4·320 "
Coffee, 2¾ oz. (I assume that this is coffee roasted and ground, and that 1 oz. gives 8 oz. of infusion, containing 0·481 gr. of nitrogen per fluid oz.),	10·582 "
Tea, 1 oz. (I estimate 48 oz. of infusion, containing 0·087 gr. of nitrogen per fluid oz.),	4·176 "
Sugar, ¼ lb.; grapes, ¾ lb.; 1½ orange (no nitrogen),	....
Total nitrogen in twenty-four hours,	781·730 grains.

analysis for uric acid on the ground that I used nitric acid for its precipitation instead of hydrochloric acid, and that the time allowed for the uric acid to separate was not sufficiently long. His experiments upon this point show conclusively that he is correct. Still, the quantities obtained by me do not appear to be small as compared with the estimates given by most authorities, and I merely gave the processes employed and the results obtained by the chemists, having had nothing to do personally with the analyses.

To obtain all of the nitrogen of the urine, Dr. Pavy might have estimated the nitrogen of the hippuric acid and of the creatine and creatinine, which would have made a difference of three or four grains of nitrogen for each day. While I cannot regard this omission as important, I must contend that the nitrogen of the fæces should have been estimated. In my experiments, the daily average of nitrogen of fæces for the five days before the walk was 21·91 grains, for the five days of the walk, 24·32 grains, and for the five days after the walk, 33·99 grains. As Dr. Pavy states, it may not be assumed that the nitrogen of the fæces resulted from any waste of tissue; still, if we are calculating accurately the ingress and the egress of nitrogen, the fæces should be taken into account, and, at least, the nitrogen of the fæces should be deducted from the nitrogen of food, as it could not be part of the nitrogen assimilated. But this, even if it be regarded as an error in Dr. Pavy's experiments, is not very important.

In the following table, which gives the main results of Dr. Pavy's observations, I have myself calculated for each period the proportion of nitrogen excreted in the urea and uric acid to the nitrogen of food:

54 THE SOURCE OF MUSCULAR POWER.

Tabular Arrangement of Results obtained for Days of Walking and Rest.—PAVY, *Effect of Prolonged Muscular Exercise upon the Urine in Relation to the Source of Muscular Power* (The Lancet, London, December 9, 1876, pp. 816, 817).

	Perkins's 24 Hours' Walk.	Perkins's Day of Rest.	First 24 Hours of Weston's 48 Hours' Walk.	Second 24 Hours of Weston's 48 Hours' Walk.	Twenty-four Hours preceding Weston's 75 Hours' Walk.	First 24 Hours of Weston's 75 Hours' Walk.	Second 24 Hours of Weston's 75 Hours' Walk.	Third 24 Hours of Weston's 75 Hours' Walk.
Body Weight [1] (with Clothing)	137 lbs.	Not given.	187¾ lbs.	133 lbs. at close of walk.	Not obtained.	137 lbs.	135 lbs.	134 lbs.
Miles walked	65·5	......	100	80·5	......	104	76	88
Nitrogen of Food	315·50 grs.	357·68 grs.	65·66 grs.	161·72 grs.	528·11 grs.	522·42 grs.	371·92 grs.	718·96 grs.
Nitrogen of Urea and Uric Acid	600·63 "	273·90 "	550·95 "	461·57 "	303·19 "	496·21 "	568·60 "	538·11 "
Per cent. of Nitrogen excreted [2] in Urea and Uric Acid	190·37	76·58	888·64	265·41	57·96	94·98	64·66	74·67

	Twenty-four Hours Rest Period after Weston's 75 Hours' Walk.	First 24 Hours of Six Day Period prior to Weston's 6 Days' Walk.	Second 24 Hours of Six Day Period prior to Weston's 6 Days' Walk.	Third 24 Hours of Six Day Period prior to Weston's 6 Days' Walk.	Fourth 24 Hours of Six Day Period prior to Weston's 6 Days' Walk.	Fifth 24 Hours of Six Day Period prior to Weston's 6 Days' Walk.	Sixth 24 Hours of Six Day Period prior to Weston's 6 Days' Walk.	First 24 Hours of Weston's 6 Days' Walk.
Body Weight (with Clothing)	Not given.	(?)	(?)	(?)	(?)	(?)	(?)	134¾ lbs.
Miles walked	......	......	......	......	......	......	......	96
Nitrogen of Food	645·11 grs.	378·29 grs.	451·14 grs.	472·29 grs.	639·95 grs.	441·15 grs.	562·60 grs.	491·60 grs.
Nitrogen of Urea and Uric Acid	284·91 "	292·70 "	301·75 "	281·54 "	363·71 "	342·59 "	387·17 "	524·59 "
Per cent of Nitrogen excreted in Urea and Uric Acid	86·41	77·37	66·88	49·02	67·86	77·65	66·43	106·67

[1], [2]. See foot-notes, p. 55.

Tabular Arrangement of Results obtained for Days of Walking and Rest—(Continued).

	Second 24 Hours of Weston's 6 Days' Walk.	Third 24 Hours of Weston's 6 Days' Walk.	Fourth 24 Hours of Weston's 6 Days' Walk.	Fifth 24 Hours of Weston's 6 Days' Walk.	Sixth 24 Hours of Weston's 6 Days' Walk.	First 24 Hours after Weston's 6 Days' Walk.	Second 24 Hours after Weston's 6 Days' Walk.
Body Weight [1] (with Clothing)	182¼ lbs.	182 7/16 lbs.	182 4/16 lbs.	181¼ lbs.	180 7/16 lbs.	Not given.	Not given.
Miles walked	77	70·5	76·5	67	63	....	....
Nitrogen of Food	826·49 grs.	759·15 grs.	547·57 grs.	790·78 grs.	614·61 grs.	Not given.	Not given.
Nitrogen of Urea and Uric Acid	582·42 "	600·29 "	508·21 "	430·06 "	468·70 "	394·40 grs.	288·89 grs.
Per cent. of Nitrogen excreted [2] in Urea and Uric Acid	70·47	79·07	91·90	56·91	76·26	Cannot be estimated.	Cannot be estimated.

	Third 24 Hours after Weston's 6 Days' Walk.	Fourth 24 Hours after Weston's 6 Days' Walk.	Fifth 24 Hours after Weston's 6 Days' Walk.	Sixth 24 Hours after Weston's 6 Days' Walk.	Daily Average for the 15 Days of Rest.	Daily average for the 12 Days of Walking.
Body Weight (with Clothing)	Not given.	Not given.	Not given.	Not given.	....	....
Miles walked	....	....	....	....	....	79·91
Nitrogen of Food	Not given.	Not given.	Not given.	Not given.	487·94 grs. (for 9 days)	556·79 grs. "
Nitrogen of Urea and Uric Acid	381·22 grs.	278·46 grs.	299·19 grs.	268·17 grs.	308·49 grs. (for 9 days); 305·42 grs. for 15 days).	527·96 "
Per cent. of Nitrogen excreted in Urea and Uric Acid	Cannot be estimated.	Cannot be estimated.	Cannot be estimated.	Cannot be estimated.	62·20 (for 9 days); cannot be estimated for 15 days.	94·82

[1] For the six days before the six days' walk, the weight is given as "without clothing," and it cannot be compared with the weights of other days, which are given as "in costume."

[2] The figures in this line were calculated by myself from the results obtained by Dr. Pavy.

I shall now endeavor to follow the process of reasoning by which Dr. Pavy draws his conclusions from the facts contained in the table just given. Dr. Pavy, at the outset, makes the following statement, the accuracy of the first part of which is sufficiently evident:

"Under any way of looking at the figures, it is evident that we have an increased elimination of nitrogen to deal with during the days of walking, which is not to be accounted for by the nitrogen ingested. We can only, therefore, refer this increase to the effect of the exercise; but is it the result—is it the expression of the action which has given rise to the power evolved? This is the question that presents itself for solution, and I will attempt to solve it by ascertaining whether the force liberated by the oxidation of muscular tissue corresponding with the nitrogen discharged is sufficient to account for the work performed."

Dr. Pavy, as is seen by the above quotation, does not fail to see that the increased elimination of nitrogen is due directly to the excessive exercise, the simple proposition which I attempted to prove by my observations in 1870. This being admitted, the question which seems to occur to his mind is how this fact can be made to accord with the idea presented as a proposition preliminary to his deductions, which is the following:

"The food of animals contains force in a latent state.

Properly regarded, food must be looked upon, not simply as so much ponderable matter, but as matter holding locked-up force. By the play of changes occurring in the body the force becomes liberated, and is manifested as muscular action, assimilative, secretory, or nutritive action, heat, etc."

He assumes the fact, which is probably correct as far as can be determined by experiments out of the body, that there is a fixed dynamic or mechanical work-value for a certain quantity of albuminous matter. By means of the calorimeter, Prof. Frankland has determined the amount of heat evolved by the combustion of such matter, and, by the formula of Mr. Joule, the equivalent of working-power of this heat may readily be calculated. These calculations are taken by Dr. Pavy as the basis for his own reasoning.[1] While all this is sufficiently definite and positive, when the application is made to the muscular work performed by the subjects of his experiments, the element of inaccuracy seems to be very great. By using the methods of Frankland and Joule, it can be shown that the oxidation of the amount of nitrogenous matter represented by the nitrogen eliminated from the body would, if such oxidation were simple combustion out of the body, produce a definite amount of heat, which would be

[1] *The Lancet*, London, December 16th, p. 849.

equal to a certain number of foot-tons of work. But, in the actual experiments upon Perkins and Weston, the definite results are to be found in the amount of work actually performed. These persons walked a certain number of miles, and this is the main experimental fact. To bring this fact into the only form in which it can be compared with the calculations of the force-value of the albuminous matter represented by the nitrogen excreted, it is necessary to reduce the miles walked to foot-pounds or foot-tons of work. The following statement of the formula by which this is done shows an inaccuracy which is inevitable and is fatal to the mathematical calculations upon which Dr. Pavy's ideas are based:

"According to Prof. Haughton, the force expended in walking on level ground may be estimated as equal to that required to raise one-twentieth of the weight of the body through the distance traversed. The number of miles walked being brought into feet, multiplying by one-twentieth of the weight of the body represented in pounds, will give the work in foot-pounds; but as so extensive a series of figures has here to be dealt with, it is convenient to reduce the result to foot-tons."

To illustrate this inaccuracy, I have only to present the following calculations for Weston's six days' walk made under Dr. Pavy's observation, and for the five days' walk observed by myself in 1870:

During the six days, according to Dr. Pavy's calculations, Weston walked four hundred and fifty miles, equivalent, according to the formula of Haughton, to an amount of work represented by 7,026·06 foot-tons.[1] This was the work actually accomplished. During this time the total nitrogen of food taken amounted to 4,030·34 grains. The loss of weight of the body during that time amounted to four and one-half pounds, or 31,500 grains. Payen's calculations give the proportion of nitrogen in fresh beef without bones as three parts in one hundred.[2] This formula, applied to the muscular tissue lost during the walk, assuming that the loss of weight was due entirely to loss of muscular substance, gives for the amount of nitrogen that would represent the loss in the weight of the body 945 grains, which, added to the nitrogen of the food, gives a total of 4,975·34 grains. According to Dr. Pavy's estimate, that one grain of nitrogen oxidized represents 2·4355 foot-tons of work, the force contained in the nitrogen of food and the nitrogen of muscular tissue wasted would amount to 12,117·44

---

[1] *The Lancet*, London, December 16, 1876, p. 849.
[2] PAYEN, *Précis théorique et pratique des substances alimentaires*, Paris, 1865, p. 488.

foot-tons. This, however, is the total work represented by the nitrogen of food and of the muscular tissue lost, and from this must be deducted the force required for "nervous action, assimilative, secretory, or nutritive action, heat, etc." This latter, as is calculated by Dr. Pavy, is represented by the nitrogen excreted during the days when no muscular work was performed. To accord with my first calculation, this must be represented by the nitrogen of food for these days of rest. Restricting this calculation to the case of Weston, and taking his six days of rest prior to the walk (for the nitrogen of food is not given for the six days after the walk), I find that the nitrogen of food for these six days amounted to 2,865·62 grains. During these six days, Weston gained two pounds and two ounces in weight,[1] or 14,875 grains, which I assume to be muscular tissue containing 446·25 grains of nitrogen. This nitrogen, which I assume, for sake of argument, went to accumulation of body-weight, is to be deducted from the nitrogen of food, the remainder being the amount of nitrogen assumed to have been expended in the force required for "nervous action, assimilative, secretory, or nutritive action, heat, etc." Making this deduction, there remain 2,419·37 grains of nitrogen

[1] *The Lancet*, London, March 11, 1876, pp. 392, 394.

of food not expended in muscular work (for there was no muscular work), and not going to increase the body-weight. The force-value of this nitrogen, according to Dr. Pavy's formula, equals 5,892·37 foot-tons. Assuming that this nitrogen represents the force required for " the internal operations going on in the system," it may be deducted from the force-value of the nitrogen of food and of the nitrogen of muscular tissue disintegrated during the six days' walk. Making this deduction, the estimated force-value of the nitrogenized food in excess of that assumed to be used in the "internal operations," with the muscular tissue consumed and not repaired, as estimated by loss of weight, is equal to 6,225·07 foot-tons. The actual work performed, reduced to foot-tons, was 7,026·06. The difference of 800·99 foot-tons, which cannot be accounted for by the nitrogenized food and muscle consumed, must be attributed to errors in the formulæ by which the calculations are made.

To render my argument complete, as far as can be done with the data at my command, I shall now apply the same methods of calculation to my own observations on Weston in 1870.

For five days before the walk, the nitrogen of food amounted to 1,697·30 grains, the force-value of which is equal to 4,133·77 foot-tons. I add 1·3 pound of weight lost, which is represented by 273 grains of nitrogen, the force-value of which is equal to 664·89 foot-tons, making a total force-value of nitrogen of food and nitrogen of muscle consumed of 4,798·66 foot-tons. From this I deduct forty-one miles walked as equal to 578·72 foot-tons of work. This leaves 4,219·94 foot-tons of work used in the "internal operations." In the calculations made from Dr. Pavy's observations on Weston for six days of rest before the walk, I found that the force used in the "internal operations" was equal to 5,892·37 foot-tons, or nearly forty per cent. more than in my observations.

I assume that the force used in these "internal operations" was nearer the normal standard during the five days before the walk than for the five days after the walk when the system was recuperating after such extraordinary muscular exertion.

During the five days of the walk of three hundred and seventeen and one-half miles, the nitrogen of food amounted to 1,173·82 grains, the force-value of which equals 2,858·79 foot-tons. The loss of weight was 3·45 pounds, representing 724·5 grains of nitrogen, having a force-value of 1,764·52 foot-tons. This, added to the force-value of the nitrogenized food, gives a total force-value of nitrogenized food and muscle consumed of 4,623·31 foot-tons. If I now deduct the force used in the "internal operations," as calculated for the five days before the walk, I have 403·37 foot-tons of force with which to accomplish a walk of three hundred and seventeen and a half miles. The actual work performed in walking three hundred and seventeen and a half miles, using Dr. Pavy's method of calculation, is equal to 4,321·33 foot-tons,[1] as is seen by the following table:

Miles walked in Five Days reduced to Foot-Tons.

	WEIGHT.	MILES WALKED.	FOOT-TONS.
First day............	116·50 lbs.	80	1,098·43
Second day..........	116·25 " (Estimated.)	48	657·64
Third day............	115·00 lbs.	92	1,246·97
Fourth day...........	114·00 "	57	765·80
Fifth day............	115·75 "	40½	552·49
Totals...........	....	317½	4,321·33

This calculation shows that the actual work performed is more than ten times the estimated force-

[1] In estimating the work, I have calculated the weight of the body without clothing. The actual work performed, according to the formula of Haughton, would be more than that which I have given; but I thought it better to calculate from my actual figures than to add an indefinite quantity for probable weight of clothing. Estimating the clothing at ten pounds, the additional work for the five days would amount to 375·09 foot-tons, making the total work equal to 4,696·42 foot-tons, 73·11 foot-tons more than the total estimated force-value of food and muscle consumed.

value of the nitrogenized food and muscle consumed, deducting that used in the "internal operations;" and, estimating for the weight of clothing carried, it is even more than the total force-value of food and muscle consumed, taking no account of the force used in circulation, respiration, etc. There must be some fatal error in the basis of calculations the results of which appear to be so far removed from the actual facts.

Again, looking at the probable force employed in the "internal operations" for the five days after the walk, I find that the nitrogen of food amounted to 2,204·65 grains. I deduct from this the five pounds gain in weight, which represents 1,050 grains of nitrogen. This leaves 1,154·65 grains of nitrogen, the force-value of which is 2,812·15 foot-tons. If I deduct from this the force represented by walking eleven miles, which is equal to 156·21 foot-tons, I have remaining, for the force used in the "internal operations," 2,655·94 foot-tons. This is rather more than one-half, and a little less than two-thirds, of the force estimated as used in the "internal operations" for the five days before the walk.

The calculations made by Dr. Pavy are entirely different in principle from those which I have just detailed. He estimates the force-value of nitrogenized matter consumed as represented, not by the nitrogen of food and muscle used, but by the nitrogen excreted. This does not appear to me to be so rational and philosophical a method as the one employed by me, for the following reasons :

1. While, in normal nutrition, there would be an exact balance between the ingress and egress of nitrogen, provided the weight of the body be uniform, in Dr. Pavy's observations for the six days of rest, from 22·35 to 50·98 per cent. of the nitrogen of food is lost and cannot be accounted for by the nitrogen excreted. It cannot be assumed, upon the principles laid down by Dr. Pavy, that the nitrogen which has thus escaped observation is actually lost or that nitrogenized matter is consumed without the evolution of force. It can only be said that its changes have escaped observation. In all experiments with regard to the ingress and egress of nitrogen with which I am acquainted, a certain proportion of nitrogen appears to be lost, or at least it is not represented in the excretions. This is probably due to imperfections in our methods of investigation.

2. The data from which my calculations were made had their starting-point in positive information with regard to the quantity of nitrogen taken in the food. This being ascertained, it seems to me much more rational to calculate the power from the

nitrogen consumed than from the nitrogen of the excretions. If it be assumed that, at the beginning of the walk, Weston was in what athletes would call "good condition," the system being free from fat as far as fat can be eliminated by training, it is rational to assume that the muscular tissue consumed, as measured by loss of weight during exercise, was used in the production of force, and the force-value of this muscular tissue consumed should be added to the force-value of the food.

Adopting the method of calculation employed by Dr. Pavy, the impossibility of accounting for the work actually performed under his observation, by calculating the force-value of matter represented by the nitrogen excreted, is more striking than is shown by the results of my method of calculating the force-value of the nitrogenized food taken. Dr. Pavy, taking all of his observations upon Perkins and Weston, estimates the amount of actual work accomplished for each day in foot-tons. He then calculates the "force-value in foot-tons of nitrogenous matter equivalent to the total urinary nitrogen eliminated." He then calculates the "force-value in foot-tons of nitrogenous matter equivalent to the nitrogen eliminated in excess of the average during rest." His remarks upon these figures are the following:

EXPERIMENTS OF DR. PAVY.

	Weight of Pedestrian in lbs.	Distance walked in Miles.	Work performed, represented in Foot-Tons.	Force-Value in Foot-Tons of Nitrogenous Matter, equivalent to the Total Urinary Nitrogen eliminated.	Force-Value in Foot-Tons of Nitrogenous Matter, equivalent to the Nitrogen eliminated in Excess of the Average during Rest.
Perkins's 24 hours' walk..............	187	65¼	1057·59	1462·88	718·98
First 24 hours of Weston's 48 hours' walk...........................	187¾	100	1623·48	1341·88	597·98
Second 24 hours of Weston's 48 hours' walk	188	80¼	1261·88	1124·15	380·80
First 24 hours of Weston's 75 hours' walk	187	104	1679·23	1208·52	464·67
Second 24 hours of Weston's 75 hours' walk	185	76	1209·21	1373·18	629·28
Third 24 hours of Weston's 75 hours' walk...........................	184	68	1310·80	1298·89	554·53
First 24 hours of Weston's 6 days' walk	184⅘	96	1525·80	1277·63	533·78
Second 24 hours of Weston's 6 days' walk...........................	132⅝	77	1205·27	1416·48	674·63
Third 24 hours of Weston's 6 days' walk	132₁₆	70¼	1097·29	1462·00	718·15
Fourth 24 hours of Weston's 6 days' walk	132₁₂	76¼	1105·19	1225·56	481·71
Fifth 24 hours of Weston's 6 days' walk	131¼	67	1085·44	1096·12	352·27
Sixth 24 hours of Weston's 6 days' walk	130₁₆	63	967·57	1141·51	897·66

" In one column is the weight of the pedestrian; in the next the distance walked; and then follow the measure of work to which these two factors correspond, the work-value of the nitrogenous matter equivalent to the total urinary nitrogen eliminated; and, lastly, the work-value of the nitrogenous matter equivalent to the urinary nitrogen excreted in excess of the average for the total days of rest. I consider that this is the column which ought to be followed, for it is clearly this which represents the disintegrated nitrogenous matter arising from the walk performed. For instance, during the days of rest there was a certain amount of disintegration of nitrogenous matter occurring as a result of the internal operations going on in the system. The disintegrated nitrogenous matter thus resulting, taking the average of all the days (fifteen) of rest, corresponded with 305·42 grains of eliminated nitrogen. It may be fairly assumed that the same, or about the same, disintegration would ensue during the days of the walk, and independently of the effects of the walk. To represent, therefore, the disintegration occurring as a result of the walk, we ought strictly to deduct that which the data show would have occurred without it.

"The above figures speak for themselves. A glance at them is sufficient to show that the force obtainable from the nitrogenous matter disintegrated is totally inadequate to supply the power for the work performed. In every case the force-value of the nitrogenous matter standing in excess of that disintegrated according to the average for the days of rest, and representing, it may be considered, the effect of the exercise, falls very far short of the power expended in walking. Upon only four occasions does it amount to more than half, and upon two it is less than a third. Even in five out of the twelve days the force-value of the total nitrogenous matter disintegrated is below the force employed."[1]

I have gone thus elaborately into a discussion of the application of what we know concerning the force-value of nitrogenized food, as calculated from the heat-units, to muscular power, for the reason that the application of the known laws of dynamics to physiological processes has lately become an important element in the study of physiology. It is to be feared, however, that physiologists are often

[1] *The Lancet*, London, December 16, 1876, pp. 849, 850.

reasoning in advance of the experimental development of facts, and theorizing upon the basis of formulæ so inaccurate, that, when they come to the consideration of millions of foot-pounds, the variations between the facts and their calculations become enormous. If such a method of study be accurate, the advantages of its application to physiology are almost incalculable. If it be shown that it is grossly erroneous, it should be abandoned until our data are more complete and definite. I cannot see how we can avoid banishing, for the present, at least, these uncertain and erroneous processes from physiological research as applied to the theories of muscular action and the source of muscular power; and I may sum up my reasons for this view in the following statements:

1. The only definite fact available in such calculations as are made by Haughton, Dr. Pavy, and others, is, "that the force derived from chemical action which will raise the temperature of a pound of water 1° Fahr. will, under another mode of manifestation, lift 772 pounds one foot high."

2. According to this proposition, if food be burned in oxygen, it will develop a definite amount of heat, which is calculated to be equivalent to a certain number of foot-pounds of force.

3. In applying this method of reasoning to the

development of force produced by the metamorphosis which nitrogenized food undergoes in the animal organism, a calculation is made of the exact force-value of food; but the processes going on within the animal organism are so complex and so imperfectly understood, that it is necessary to establish a certain relation between the calculated force-value of the food and the actual force exerted under certain conditions. Experimental attempts to establish such a relation have signally failed.

4. It cannot be estimated, with any approach to scientific accuracy, how much force is exerted by the heart. One observer makes this force nearly double that estimated by another.[1] Dr. Haughton bases his estimate upon the proposition that the left ventricle discharges three ounces of blood at each contraction. There is good physiological authority for the opinion that the quantity discharged at each pulsation is from five to six ounces. An error in this estimate, when it is remembered that this error must be multiplied by over 100,000 beats of the

---

[1] I may illustrate the differences in the estimates of different observers by stating that Haughton estimates the "total daily work of both ventricles" as equal to 124·208 foot-tons, or 621·04 foot-tons for five days. Letheby calculates the daily work of the heart (seventy-five beats in a minute) as equal to 223·23 foot-tons, or 1,116·15 foot-tons for five days.—(HAUGHTON, *Principles of Animal Mechanics*, London, 1873, p. 145; and LETHEBY, *On Food*, New York, 1872, p. 96.)

heart in twenty-four hours, it is evident, would be fatal to the accuracy of any calculations.

5. It is impossible to arrive at any reasonably accurate estimate of the force exerted in the movements of respiration.

6. There can be no experimental estimate, unless it be based upon calculations derived from the heat-units of food, of the amount of heat generated within the body under such conditions as those presented in the observations now in question. It has been possible to estimate approximately the actual production of heat by the body in a state of rest, by noting the elevation of temperature of a definite quantity of water in a bath; but, during violent exercise in the air, it seems impossible to calculate the radiation of heat from the surface with any degree of accuracy, and the production of heat under such conditions must remain, for the present, at least, an unknown quantity.

7. It is absolutely necessary to have an accurate estimate of the force used in circulation, respiration, and in the production of animal heat, so that this force may be deducted from the force-value of food, in estimating the force used in ordinary muscular work.

8. If the calculations of the actual force used in muscular work be made from persons walking, there

must be some accurate method of reducing this force, for purposes of comparison, to foot-pounds or foot-tons. I cannot accept the proposition that the work accomplished in walking on a level is equal to raising one-twentieth of the weight of the body the distance walked, as is done by Dr. Pavy in his calculations. This is taken from Haughton's calculations, in which the velocity was three miles per hour,[1] while the actual velocity in Dr. Pavy's observations must have been between four and one-half and five miles per hour.[2] It would seem sufficiently evident, from the propositions which I have just stated, that Dr. Pavy's estimates must be wanting in accuracy, as nearly all of the elements entering into his calculations are necessarily very indefinite. In calculating from his observations upon Weston, the facts of which I assume to be reasonably accurate, over eleven per cent. of the work actually accomplished cannot be accounted for either by the force-value of nitrogenized food or by the force-value of muscular substance consumed and not repaired. I assume, also, that the facts obtained in my observations are reasonably accurate. In these, the actual

[1] HAUGHTON, *Principles of Animal Mechanics*, London, 1873, p. 57.

[2] In the observations which I made upon Weston in 1870, the rate of walking for the five days averaged between four and one-half and five miles per hour.

work accomplished in a walk of three hundred and seventeen and one-half miles in five days was more than ten times the estimated force-value of nitrogenized food and muscle consumed, deducting the force used in circulation, respiration, etc.

Assuming that force is evolved only as a consequence of the metamorphosis of matter, and that it is only food and muscle (and perhaps fat) that can undergo changes producing force in muscular work, when calculations show a large excess of force actually produced over the estimated force-value of nitrogenized food and of muscle consumed, it is evident that there must be a serious error, either in the measurement of the force produced, or in the calculations of the force-value of matters consumed, or in both. There is, in all probability, an error in both. The reduction of level miles walked to foot-tons is inaccurate; an accurate estimation of the force used in circulation, respiration, etc., seems at present impossible; and the assumption that the force-value of nitrogenized food, calculated by reducing heat-units developed by burning the food in oxygen to foot-tons, can be applied absolutely to the changes which food undergoes in the human body has no argument in its favor drawn from experimental facts. Still, no one can say that matter can be actually destroyed, that matter can undergo certain chemical changes without the development of force, or of heat which represents force, or that force can be developed in the body without some change in matter. My only argument is that purely physical laws cannot as yet be applied absolutely to the operations of the living organism.

Putting aside for the present theoretical considerations, I propose to discuss what can be learned from the facts developed by the observations made by Dr. Pavy upon Perkins and Weston in 1876, and by myself upon Weston in 1870.

Dr. Pavy's first observation is upon Perkins, who walked sixty-five and one-half miles in twenty-four hours. During this time, he took 315·50 grains of nitrogen in food, and excreted 600·63 grains of nitrogen in urea and uric acid. For every 100 parts of nitrogen of food, he excreted 190·37 parts of nitrogen in the urine. A number of days after, Dr. Pavy noted the amount of nitrogen consumed and excreted by Perkins for a day of rest. He took in 357·68 grains with the food, and discharged 273·90 grains in the urine, or 76·57 parts for every 100 parts of nitrogen of food. As far as this observation goes, it shows that the work of walking sixty-five and one-half miles in a day increased the proportionate elimination of nitrogen in the urine nearly two hundred and fifty per cent. No account is

EXPERIMENTS OF DR. PAVY.

given of the variations in weight produced by the walk. The greatest proportionate increase in the excretion of nitrogen which I observed for any one day in Weston was 224·32 per cent.

Dr. Pavy's observations upon Weston, during a walk of forty-eight hours, showed the following: During the first twenty-four hours, walking one hundred miles, Weston took 65·68 grains of nitrogen in his food (an exceptionally small quantity), and excreted 550·95 grains in the urine. For every 100 parts of nitrogen of food he excreted 838·84 parts of nitrogen by the urine. During the second twenty-four hours, the quantity of nitrogen of food was still quite small (161·72 grains), and the nitrogen of the urine amounted to 461·57 grains, or 285·41 parts for every 100 parts of nitrogen of food. Comparing this with the average proportion of nitrogen excreted for eight days of rest (Weston), which is 60·90 parts excreted by the urine for every 100 parts in the food, the figures show an enormous proportionate increase of excreted nitrogen, which can be attributed only to the excessive muscular work. The unusually small amount of nitrogen taken in the food was evidently incapable of supplying the waste of tissue, which amounted to a loss of body-weight of four and three-fourths lbs.; but it must be evident to any one that the excessive proportionate excretion of nitrogen was produced, in this instance, at least, by the muscular work, and that it bears a certain relation to the amount of muscular tissue consumed. This observation is very instructive, as showing the effects of excessive muscular work with a great deficiency in the supply of reparative matter. I shall discuss this point more fully in connection with my own observations, in which, as I knew, Weston was in what would be called " good condition." In Dr. Pavy's observations, I do not know the amount of fat that may have been lost during the walk.

The next observations by Dr. Pavy were upon Weston in a walk of seventy-five hours. For twenty-four hours of rest preceding this walk, with 523·11 grains of nitrogen of food (a little above the usual quantity at rest), he excreted a percentage of 57·96 of nitrogen in the urine. On the first day of the walk, with 522·42 grains of nitrogen of food, he excreted 94·98 per cent. in the urine. On the second day, with 871·92 grains of nitrogen of food (the largest quantity noted for any one day), he excreted 64·66 per cent. in the urine. On the third day, with 713·96 grains of nitrogen of food, he excreted 74·67 per cent. in the urine. On a day of rest immediately following this walk, when he was recuperating from the effects of the exertion, with

645·11 grains of nitrogen of food, he excreted only 36·41 per cent. of nitrogen in the urine, the lowest proportionate excretion observed for any one day. These figures show a very large proportionate increase in the excretion of nitrogen during the walk, which must be attributed to the muscular work; but the proportionate increase is not so great as in the forty-eight hours' walk, for the reason that the supply of nitrogen in the food appeared to be sufficient to repair the waste of tissue—if it be assumed that the work produced waste of tissue, as it must have done in the walk of forty-eight hours. The loss of weight was three pounds, but it is impossible to say how much of this loss was fat or water.

Dr. Pavy's next observations were upon Weston's six days' walk. During six days of rest just before the walk, the average nitrogen of food daily was 477·60 grains, the average nitrogen excreted in the urine daily was 319·91, or 66·98 parts for every 100 parts of nitrogen of food.

During the six days of the walk, the total distance walked being four hundred and fifty miles, the average nitrogen of food daily was 671·72 grains. This is a very large quantity, nearly three times the average obtained by me for a five days' walk. The average daily excretion of nitrogen in the urine was 521·54 grains. This is also a very large quantity, fifty-four per cent. more than I obtained for a five days' walk. For every 100 parts of nitrogen of food during this six days' walk, Weston excreted 77·64 parts in the urine, against 66·98 parts during six days of rest.

No calculations can be made for the six days of rest immediately after the walk, as the nitrogen of food was not estimated. It is stated that the diet " comprised a generous daily mixed allowance of animal and vegetable food." The average nitrogen excreted daily in the urine for the six days of rest after the walk was 308·30 grains, a little less than the average quantity for the six days of rest before the walk. It is probable that, while recuperating for six days after the walk, the quantity of food was greater than for the six days before the walk.

During the six days' walk, there was nothing notable except a large proportionate excretion of nitrogen on the first day (106·67 per cent.), when he walked ninety-six miles, which was the longest distance walked in any one day.

An important criticism which I have to make with regard to all of the observations of Dr. Pavy is that he took no account of the nitrogen discharged in the fæces, which should certainly have been estimated as nitrogen discharged from the body and deducted from the nitrogen assumed to have been consumed.

As regards the observations on Weston during the six days' walk, the same remarks apply as those made for the seventy-five hours' walk. The figures show a large proportionate increase in the excretion of nitrogen during the walk over the excretion during the six days of rest immediately preceding. This, as Dr. Pavy admits, is only to be attributed to the influence of the muscular work. It is a noticeable fact, however, that the quantity of nitrogen taken in the food during the walk was enormous; and the food undoubtedly supplied the excessive waste of tissue produced by the extraordinary exertion to a much greater extent than during the five days' walk observed by me in 1870. It probably prevented a complete breaking down, such as occurred on the fourth day of my observations. On this day, the nitrogen of food amounted to only 144·70 grains; the excretion by the urine was 324·59 grains, or 224·32 parts for every 100 parts of nitrogen of food; the proportionate excretion for the three days before had amounted to a daily average of 163·86 per cent., and the loss of weight for the four days had been 5·2 pounds. The muscular system then seemed to be incapable of farther work. Weston became almost blind, and he was taken from the track and supported to his room. If Weston had been able to supply his waste of tissue by food, this collapse of muscular power might not have occurred.

It is evident that, in Dr. Pavy's observations, the proportionate discharge of nitrogen during the six days' walk was less than in mine, for the reason that the nitrogen of food was very much greater, being more than double the quantity that I obtained; but it is evident, also, that in my experiments Weston suffered from want of food, and the increased quantity of food under Dr. Pavy's observation was useful in repairing the muscular tissue.

I have separated the different series of observations made by Dr. Pavy, for the reason that one series was made upon Perkins and the others upon Weston; and also because the conditions of alimentation in the different series of observations upon Weston presented great variations, which were particularly noticeable in the forty-eight hours' walk.

Important changes occurred in the quantities of inorganic matters in the urine during the extraordinary exertions made under Dr. Pavy's observation. Dr. Pavy's results, in this regard, were nearly the same as mine. The physiological significance of these changes is not very clear, and I refrain from discussing them, for the reason that they seem to have no very distinct relations to the source of muscular power.

My own Experiments, made in 1870, and published in 1871.[1]

In discussing my own experiments, which I can now do more satisfactorily than before, by reason of the confirmatory results obtained by Dr. Pavy, I feel that I am justified in claiming priority in the method of investigating the influence of exercise upon the excretion of nitrogen by comparing the nitrogen eliminated with the nitrogen of food.

I endeavored to make my experimental data complete and accurate, and to gather and group my facts with entire freedom from theoretical bias. I was with Weston for the entire fifteen days, and had an assistant who never left him for an instant and even slept with him at night. Every article of food and drink was weighed or measured, and the amount of nitrogen estimated. No part of the urine or fæces was lost. No accident happened, the weights were all taken naked, and the subject of the experiments was under observation every instant. The chemical analyses were all made by a chemist, who had no idea of the ultimate bearing of the re-

[1] *New York Medical Journal*, June, 1871.

sults which he was to present to me for tabulation. Probably the only error in these analyses was in the estimation of the uric acid. This was not so considerable as to affect the conditions in any way, and the same error extended through all the examinations, so that it could not modify the comparative results. In my calculations of average percentages, I made an error of taking the average of the percentages instead of the percentage of the averages. This I corrected in *The Journal of Anatomy and Physiology*, Cambridge and London, October, 1876. Dr. Pavy takes me to task for this error in *The Lancet*, London, December 23, 1876, some time after it had been corrected; but my correction of the figures shows that it in no way affected my general conclusions.

In the following table, which gives the most important results of my experiments, the error alluded to above has been corrected. I have also left out the estimates of the nitrogen of the fæces and added the nitrogen of the uric acid, so as to be able to make a more accurate comparison of my results with those obtained in England:

Tabular Arrangement of Results obtained for Days of Walking and Rest.—FLINT, JR., *Physiological Effects of Severe and Prolonged Muscular Exercise* (New York Medical Journal, June, 1871).

	First 24 Hours of 5 Day Period prior to Weston's 5 Days' Walk.	Second 24 Hours of 5 Day Period prior to Weston's 5 Days' Walk.	Third 24 Hours of 5 Day Period prior to Weston's 5 Days' Walk.	Fourth 24 Hours of 5 Day Period prior to Weston's 5 Days' Walk.	Fifth 24 Hours of 5 Day Period prior to Weston's 5 Days' Walk.	First 24 Hours of Weston's 5 Days' Walk.	Second 24 Hours of Weston's 5 Days' Walk.	Third 24 Hours of Weston's 5 Days' Walk.	Fourth 24 Hours of Weston's 5 Days' Walk.
Body Weight (without Clothing)	120·5 lbs.	121·25 lbs.	120 lbs.	118·5 lbs.	119·2 lbs.	116·5 lbs.	116·25 lbs.	(Estimate) 115 lbs.	114 lbs.
Miles walked	15	5	5	15	1	80	48	92	57
Nitrogen of Food	361·22 grs.	288·85 grs.	272·27 grs.	335·01 grs.	440·48 grs.	151·55 grs.	265·92 grs.	228·61 grs.	144·70 grs.
Nitrogen of Urea and Uric Acid	304·55 "	276·84 "	305·06 "	268·87 "	299·31 "	381·44 "	328·05 "	399·16 "	324·59 "
Per cent. of Nitrogen excreted in Urea and Uric Acid	84·81	96·01	112·05	84·78	67·96	218·70	128·86	174·60	224·32

Tabular Arrangement obtained for Days of Walking and Rest (Continued).

	Fifth 24 Hours of Weston's 5 Days' Walk.	First 24 Hours after Weston's 5 Days' Walk.	Second 24 Hours after Weston's 5 Days' Walk.	Third 24 Hours after Weston's 5 Days' Walk.	Fourth 24 Hours after Weston's 5 Days' Walk.	Fifth 24 Hours after Weston's 5 Days' Walk.	Daily Average for the 5 Days of Rest prior to Weston's 5 Days' Walk.	Daily Average for the 5 Days of Rest after Weston's 5 Days' Walk.	Daily Average for the 10 Days of Rest.	Daily Average for the 5 Days of Walking.
Body Weight (without Clothing)	115·75 lbs.	118 lbs.	120·25 lbs.	120·25 lbs.	123·5 lbs.	120·75 lbs.	119·89 lbs.	120·55 lbs.	120·22 lbs.	115·5 lbs.
Miles walked	40·5	2	2	2	2	3	8·2	2·2	5·2	63·5
Nitrogen of Food	338·04 grs.	385·65 grs.	499·10 grs.	394·58 grs.	641·71 grs.	283·35 grs.	339·46 grs.	440·98 grs.	390·19 grs.	234·76 grs.
Nitrogen of Urea and Uric Acid	306·80 "	277·00 "	384·44 "	358·78 "	348·19 "	319·79 "	293·93 "	339·64 "	316·78 "	338·01 "
Per cent. of Nitrogen excreted in Urea and Uric Acid	90·09	71·82	67·01	90·87	54·26	184·08	86·58	77·08	81·18	143·98

After the full discussion which I have given of Dr. Pavy's results, in which I have referred more or less fully to my own observations, a very brief general statement of the main facts embodied in the table will suffice.

My observations were made for three periods of five days each, before, during, and after the walk of three hundred and seventeen and a half miles in five consecutive days.

For the five days immediately preceding the walk, the average daily quantity of nitrogen ingested was 339·46 grains. The average daily quantity of nitrogen excreted in the urine was 293·93 grains. The proportionate excretion of nitrogen was 86.58 parts for every 100 parts of nitrogen of food.

At the beginning of the walk, Weston appeared to be in good condition, with no superfluous fat, and his weight, without clothing, was 119·2 lbs. During the five days, he walked three hundred and seventeen and a half miles. His daily average of nitrogen of food amounted to 234·76 grains. His daily average excretion of nitrogen in the urine was 338·01 grains. The proportionate excretion of nitrogen was 143·98 parts for every 100 parts of nitrogen of food. It is evident, therefore, that the nitrogen of food did not supply the waste of the nitrogenized constituents of the body, as it may be supposed to have done in Dr. Pavy's observations.

The most notable event in the course of the five days' walk was what appeared to be a total collapse of muscular and nervous power, to which I have already referred, occurring on the fourth day. I quote the following account of this from my original article, in which I calculated the nitrogen of the urea and the fæces [1] and took no account of the uric acid:

"On the third day Weston walked ninety-two miles, with thirty minutes' sleep. The entire quantity of nitrogen of the urea (no fæces were passed) was enormous, amounting to 397·58 grains, representing 851·95 grains of urea, by far the largest amount discharged for any one of the five days. This corresponded to the greatest amount of muscular exertion, a fact which is very significant. The nitrogen of the food was slightly diminished, amounting to 228·61 grains. For every 100 parts of nitrogen introduced, there were discharged 173·91 parts. This excessive discharge of nitrogen can only be attributed to the muscular exertion. On that day, Weston took six pints of strong coffee, which, if it had any effect, would have diminished the elimination of urea.

"On the fourth day Weston walked fifty-seven miles, with one hour of sleep. The nitrogen of the urea and fæces amounted to 348·53 grains. The nitrogen of the food was on

---

[1] The nitrogen of the fæces for the first four days of the walk amounted to 95·42 grains.

this day diminished to the minimum, amounting to only 144·70 grains. For every 100 parts of nitrogen introduced, there were discharged 240·86 parts, the largest excess observed during the five days.

"At 10.30 P. M., on this day, Weston broke down completely. He could not see the track, and was taken staggering to his room, having reached, apparently, the limit of his endurance. His condition at that time, as shown by the records, was as follows: He had lost in weight 5·2 lbs., being reduced from 119·2 lbs. to 114 lbs. He had taken a daily average of 197·70 grains of nitrogen in his food, while walking an average of sixty-nine and a quarter miles *per diem*, with an average of sleep in the twenty-four hours of one hour and forty-four minutes for four days. His daily average of nitrogen should have been 310 grains, not allowing for an increased quantity demanded to supply the waste engendered by his excessive muscular exertion. He had discharged for every 100 parts of nitrogen introduced, a daily average of 186·37 parts for four days. The calculations, as well as the general condition of the system, show that the period had probably arrived when repair of the muscular system had become absolutely necessary.

"On the fifth day, after nine hours and twenty-six minutes of sleep, the system reacted completely, and Weston walked forty and a half miles. The nitrogen of the urea and fæces was 332·77 grains. The nitrogen of the food was increased one hundred and sixty-five per cent., amounting to 383·04 grains. For every 100 parts of nitrogen of food there were discharged 84·27 parts. The absolute quantity of nitrogen discharged was still very great, but its proportion to the nitrogen taken in was reduced by the great quantity in the food.

"On this day, when there was an apparent reaction after the complete prostration of the fourth day, the system seemed to appropriate nitrogen, as it were, with avidity, to repair the impoverished muscular tissue. The weight was increased on this day by one and three-quarters lbs."

There is one important series of calculations made by me for the five days of the walk in 1870, which I cannot make for any of Dr. Pavy's experiments, for the reason that I do not know the physical "condition" Weston was in when the observations were made.[1] At the beginning of the walk

[1] One hour and a half before Weston began his six days' walk, he weighed 123·75 lbs., without clothing (*The Lancet*, March 18, 1876, p. 430), which was 4·55 lbs. more than he weighed in 1870, at the beginning of his five days' walk. I am unable to determine what part, if any, of this excess of weight was fat. During the six days' walk he excreted 3,129·27 grains of nitrogen in the urine, which was 1,101·21 grains more than the daily average for the five days' walk, multiplied by six. During the same time, he took 4,030·34 grains of nitrogen in the food, which was nearly three times the average for the five days' walk, multiplied by six. At the end of the six days' walk, he had lost 5 lbs. 5 oz., having been reduced in weight to 118 lbs. 7 oz. (*The Lancet*, March 18, 1876, p. 431), or only about 12 oz. less than he weighed in 1870, when he began his five days' walk. This loss in weight occurred in the six days' walk, when he had been taking in much more nitrogen than he excreted. The uncertainty with regard to the amount of fat which he may have lost, the enormous amount of food taken, and the fact that, while losing largely in weight, he actually took in more nitrogen than he excreted, render it impossible for me to make any calculations with regard to the relations between the excretion of nitrogen and the loss of body-weight for the six days' walk.

made in 1870, thirty minutes before starting, he weighed 119·2 lbs., without clothing. There was no superfluous fat, and he started apparently with his system in such a condition that nearly all the variations in weight could be attributed to a loss of muscular substance not repaired by food. This is an important consideration, for loss of fat could not be calculated from any data furnished by an analysis of the food and excretions for nitrogen. The loss of water might become a slightly disturbing element, for it is well known that three or four pounds may be lost in a Turkish bath, without any muscular exertion; but I assume that this element of error cannot have been very considerable, as Weston took all the liquids he desired during the five days, and it is probable that he promptly supplied in this way the water lost by cutaneous transpiration. When a person loses two or three pounds in a Turkish bath, the water is generally restored by drink in the course of a few hours.

At the end of the five days' walk, the weight had been reduced from 119·20 to 115·75 lbs., showing a loss of 3·45 lbs. According to Payen, three parts of nitrogen represent one hundred parts of fresh muscular tissue.[1] The total quantity of nitro-

---

[1] PAYEN, *op. cit.*, p. 488.

gen contained in the urea, uric acid, and fæces[1] discharged during these five days was 1,811·62 grains. The total nitrogen of food during the same period amounted to 1,173·82 grains, giving an excess of 637·80 grains of nitrogen discharged over the nitrogen of food. The 637·80 grains of nitrogen, according to Payen's formula, would represent 21,260·00 grains, or 3·037 lbs. of muscular tissue. The actual loss of muscular tissue was 3·45 lbs., and the loss unaccounted for, amounting to 0·413 of a lb., is very small. It might be fat or water, or the difference might be due to inaccuracies in the estimates of the nitrogen of food, which of necessity were approximative.

It is thus seen that my observations on Weston's walk of three hundred and seventeen and a half miles in five days show not only that the excessive exercise increased the amount of nitrogen discharged over the nitrogen taken in with the food, but that the excess of the nitrogen discharged over the nitrogen of the food, calculated as representing muscular tissue destroyed, was nearly equal to the actual loss of weight during the walk. This calculation is made upon the assumption that when Weston began his walk nearly all the fat had been elim-

---

[1] The nitrogen of the fæces for the five days amounted to 121·58 grains.

inated from the body, and that the loss of weight involved the muscular system almost exclusively.

The observations during the five days after the walk, with very little exercise, show that the gain in weight was five pounds, and that the nitrogen of food was 2,204·64 grains, while the nitrogen of the urea, uric acid, and fæces, was 1,868·13 grains, giving an excess of nitrogen of food of 336·51 grains. This excess of nitrogen would build up 1·602 of a pound of muscular tissue. It is fair to suppose that, under a liberal diet after such severe exertion and without exercise for five days, the remaining 3·398 lbs. of gain in weight was probably fat and not muscle.

Conclusions.

I. While the various elements of nitrogenized food burned in oxygen out of the body will produce a definite amount of heat which may be calculated as equivalent to a definite number of foot-pounds of force, the application of this law to the changes which food or certain of the constituents of the body undergo in the living organism is uncertain and unsatisfactory, for the following reasons:

(*a.*) There is no proof that the elements of food undergo the same changes in the living body as when burned in oxygen, or that definite amounts of heat or force are necessarily manifested by their metamorphoses in such a way that they can be accurately measured.

(*b.*) Assuming that the elements of food contain a definite amount of locked-up force, to measure the part of this force which is expended in muscular work, it is indispensable to be able to estimate accurately the force used in circulation, respiration, and the various nutritive processes, and to measure the heat evolved which maintains the standard animal temperature and which compensates the heat lost by evaporation from the general surface. It does not seem that any accurate idea can be formed of the amount of force used in circulation and respiration, and the estimates made by different observers of authority present variations sometimes of more than one hundred per cent. Such estimates are usually made in view of some dynamic theory, and they are based upon physiological data which are necessarily uncertain and subject to wide and frequent variations. No approximate estimate, even, can be made of the actual amount of heat produced within the living organism, except, perhaps, during a condition of nearly absolute muscular rest. The only way in which this could be done would be to deduct the force used in muscular work, circulation, respiration, and the nutritive processes from the heat or force-value of the food. These elements of the question being uncertain, an accurate estimate of the heat produced becomes impossible, as, at the best, the only definite quantity in the problem is the total heat or force-value of food.

(c.) To compare an amount of muscular work actually performed with the estimated force-value of food, apart from the impossibility of arriving at an accurate estimate of the amount of food consumed in circulation, respiration, the nutritive processes, and the production of heat, which is a necessary element in the problem, the work actually performed in walking a certain distance must be reduced to foot-pounds or foot-tons. The formula for this is so uncertain that no such reduction can be made which can be assumed to be even approximatively correct.

II. The method of calculating the possible amount of the force of which the body is capable, by using as the sole basis for this calculation the force-value of food, must be abandoned until the various necessary elements of the problem can be made sufficiently accurate to accord with the results of experiments upon the living body. Until that time arrives, physiologists should rely upon the positive results obtained by experiments rather than upon calculations made from uncertain data and under the influence of special theories. In case of fatal disagreement between any theory and definite experimental facts, the theory must be abandoned, provided the facts be incontestable.

III. Experiments show that the estimated force-value of food, after deducting the estimated amount used in circulation, respiration, the nutritive processes, and in the production of heat, will sometimes account for a small fraction only of muscular work actually performed, this work being reduced to foot-tons by the uncertain process to which I have already alluded. The errors in these calculations are manifestly so considerable that the results seem to be of little value, while the experimental fact that a certain amount of work has been accomplished must remain.

IV. It must be admitted that, under ordinary and normal conditions of diet and muscular exercise, the weight of the body being uniform, the ingress and egress of matter necessarily balance each other. If this balance be disturbed by diminishing the supply of food below the requirements of the system for its nutrition and for muscular work, the body necessarily loses weight, a certain portion of its constituent parts being consumed and not repaired. If the balance be disturbed by increasing the muscular work to the maximum of endurance and beyond the possibility of complete repair by food, the body loses weight. The probable source of muscular power may be most easily and satisfactorily studied by disturbing the balance between consumption and repair by increasing the work. In this, it is rational to assume that the processes of physiological wear of the tissues are not modified in kind, but simply in degree of activity.

V. Experiments show that excessive and prolonged muscular exercise may increase the waste or wear of certain of the constituents of the body to such a degree that this wear is not repaired by food. Under these conditions, there is an increased discharge of nitrogen, particularly in the urine. This waste of tissue may be repaired if food can be assimilated in sufficient quantity, but in my experiments it was not repaired. The most important question to determine experimentally in this connection is with regard to the influence of excessive and prolonged muscular exercise upon the excretion of nitrogen. It is shown experimentally that such exercise always increases the excretion of nitrogen to a very marked degree, under normal conditions of alimentation; but the proportionate quantity to the nitrogen of food is great when the nitrogen of food remains the same as at rest, and is not so great, naturally, when the nitrogen of food is increased. In the latter case, the excessive waste of the tissues is in part, or it may be wholly, repaired by the increased quantity of food. Experiments upon excessive exertion with a non-nitrogenous diet are made under conditions of the system that are not physiological; and the want of nitrogen in the food in such observations satisfactorily accounts for the diminished excretion of nitrogen.

VI. By systematic exercise of the general muscular system or of particular muscles, with proper intervals of repose for repair and growth, muscles may be developed in size, hardness, power, and endurance. The only reasonable theory that can be offered in explanation of this process is the following: While exercise increases the activity of disassimilation of the muscular substance, a necessary accompaniment of this is an increased activity in the circulation in the muscles, for the purpose of removing the products of their physiological wear. This increased activity of the circulation is attended with an increased activity of the nutritive processes, provided the supply of nutriment be sufficient, and provided, also, that the exercise be succeeded by proper periods of rest. It is in this way only that we can comprehend the process of development of muscles by training; the conditions in training being exercise, rest following the exercise, and appropriate alimentation, the food furnishing nitrogenized matters to supply the waste of the nitrogenized parts of the tissues. This theory involves the idea that muscular work consumes a certain part of the muscular substance, which is repaired by food. The theory that the muscles simply transform the elements of food into force directly, these elements not becoming at any time a part of the muscular substance, is not in accordance with the facts known with regard to training.

VII. All that is known with regard to the nutrition and disassimilation of muscles during ordinary or extraordinary work teaches that such work is always attended with destruction of muscular substance, which may not be completely repaired by food, according to the amount of work performed and the quantity and kind of alimentation.

VIII. In my experiments upon a man walking three hundred and seventeen and one-half miles in five consecutive days, who at the beginning of the five days had no superfluous fat, the loss of weight was actually 3·45 lbs., while the total amount of nitrogen discharged from the body in the urine and fæces in excess of the nitrogen of food taken for these five days, assuming that three parts of nitrogen represent one hundred parts of muscular substance, as has been shown by analysis to be the fact, was equivalent to 3·037 lbs. of muscular substance.

This close correspondence between the actual loss of weight and the loss that should have occurred, as deduced from a calculation of the nitrogen discharged in excess of the nitrogen of food, seems to show very clearly that, during these five days of excessive muscular work, a certain amount of muscular substance was consumed which had not been repaired, and that this loss could be calculated with reasonable accuracy from the excess of nitrogen excreted.

IX. Finally, experiments upon the human subject show that the direct source of muscular power is to be looked for in the muscular system itself. The exercise of muscular power immediately involves the destruction of a certain amount of muscular substance, of which the nitrogen excreted is a measure. Indirectly, nitrogenized food is a source of power, as, by its assimilation by the muscular tissue, it repairs the waste and develops the capacity for work; but food is not directly converted into force in the living body nor is it a source of muscular power, except that it maintains the muscular system in a proper condition for work. In ordinary daily muscular work, which may be continued indefinitely, except as it is restricted by the conditions of nutrition and the limits of age, the loss of muscular substance produced by work is balanced by the assimilation of alimentary matters. A condition of the existence of the muscular tissue, however, is that it cannot be absolutely stationary, and that disassimilation must go on to a certain extent, even if no work be done. This loss must be repaired by food to maintain life. A similar condition of existence applies to every highly-organized part of the body and marks a broad distinction between a living organism and an artificially constructed machine, which latter can exert no motive power of itself and can develop no force that is not supplied artificially by the consumption of fuel or otherwise.

APPENDIX.

In discussing the various observations that have been made upon the human subject with regard to the influence of muscular exercise upon the excretion of nitrogen, I have confined myself to the question of the relations between nitrogenized food and muscular power because, in all such observations, no account has been taken of the elimination of carbonic acid.

In my experiments, made in 1870, the quantities of non-nitrogenized food were carefully noted; and it may be interesting to speculate with regard to the possible influence of such matters upon the production of heat and work. I must premise, however, what I shall have to say upon this point, with the statement that I cannot accept the estimates given of the force used in circulation, respiration, and the production of animal heat, as even approximatively correct. With this reservation, I propose to discuss these estimates, and see what possible relation they bear to food, including non-nitrogenized as well as nitrogenized matters.

Weston walked, under my observation, three hundred and seventeen and one-half miles in five consecutive days. Making my calculations according to the method employed by Dr. Pavy, the force-value of his nitrogenized food, during these five days, was 2,858·79 foot-tons. The force-value of his loss of weight, calculated as muscular tissue, was 1,764·52 foot-tons. During the five days, he took non-nitrogenized food which represented 19,521·41 heat-units.[1] All these represent the sum of the sources of power and heat, with which Weston was to accomplish his walk of three hundred and seventeen and one-half miles and maintain circulation, respiration, animal temperature, etc.

Non-nitrogenized Food taken by Weston during his Five Days' Walk.

Articles of Food.	1st Day.	2d Day.	3d Day.	4th Day.	5th Day.	Total Oz.	Total in Grains.	Heat-Units.
Milk....	5·66 oz.	5·66 oz.	6·18 oz.	8·75 oz.	9·78 oz.	36·03	15,763·125	2,585·152
Bread...	1·25 "	10·50 "	1·50 "	6·62 "	9·00 "	28·87	12,630·625	6,972·105
Oatmeal	....	....	6·78 "	7·92 "	3·39 "	18·09	988·750	998·637
Potatoes	....	2·00 "	....	....	4·00 "	6·20	2,712·500	694·400
Butter..	2·63 "	0·50 "	0·50 "	....	1·25 "	4·88	2,185·000	3,968·160
Sugar...	1·63 "	1·75 "	2·00 "	3·62 "	2·37 "	11·37	4,974·875	4,282·987
Grand total of heat-units...								19,521·411

I have calculated the heat-units from Letheby's table (*On Food*, pp. 94, 95). In the table above, I have given the amounts of oatmeal-gruel taken, and I have estimated two ounces of oatmeal for a pint of gruel. In my calculations of the force-value of nitrogenized food, I have already estimated milk, bread, oatmeal, potatoes, and butter, taking the proportion of nitrogen for each of these articles. I have not included sugar before in any of my calculations.

APPENDIX. 101

The walk of three hundred and seventeen and one-half miles, according to Dr. Pavy's calculation, was equal to 4,321·33 foot-tons of work. According to Letheby, the force expended daily in circulation and respiration amounts to about 600,000 foot-pounds, or 3,000,000 foot-pounds (1,339·29 foot-tons [1]) in five days. Direct observations have shown that the production of heat per pound-weight of the body per hour, in a state of rest, equals 1·283 heat-units.[2] This gives 30·8 heat-units per pound-weight of the body for twenty-four hours, and 3,557·4 heat-units daily for 115·5 lbs. (Weston's average weight for the five days), and 17,787 heat-units for five days. The heat-units represented by the non-nitrogenized food taken during these five days amounted to 19,521·41. Deducting the estimated heat produced by the body, we have remaining an excess of 1,734·41 heat-units, which can be calculated as equal to 597·75 foot-tons of force. These calculations show the following force-value represented by all the food and the loss of weight of the body, exclusive of the part of the non-nitrogenized food used in the production of 17,787 heat-units:

---

[1] LETHEBY, *On Food*, New York, 1872, p. 96.
[2] DALTON, *Human Physiology*, Philadelphia, 1875, p. 302.

I calculate the heat-unit as the quantity of heat required to raise one pound of water from 0° to 1° Fahr. and have reduced the calculations from kilogrammes to pounds and from centigrade to Fahrenheit degrees of the thermometer.

Force-value of nitrogenized food...............	2,858·79 foot-tons.
" " loss of weight of the body......	1,764·52 "
" " non-nitrogenized food (excess)..	597·75 "
Total.............................	5,221·06 "
Deduct the estimated force used in circulation and respiration.........................	1,339·29 "
Force remaining for muscular work...........	3,881·77 "

The actual work represented by walking three hundred and seventeen and one-half miles is estimated at 4,321·33 foot-tons. This leaves 439·56 foot-tons of work which cannot be accounted for in any way, according to the estimates of the observers whom I have quoted, leaving a deficiency of a little more than ten per cent. These calculations show the fallacy of such estimates and the impossibility of accounting for muscular work actually performed, even when we include the heat-value and the force-value of non-nitrogenized food. The estimates of the force used in circulation and respiration and of the heat produced by the body are all calculated for a condition of rest. It is well known, however, that such unusual violent exertion as was made by Weston during his five days' walk would necessarily increase the labor of the heart and respiratory muscles and also produce a very much greater amount of heat than during rest. This would give a much greater deficiency than is shown by the estimates I have made.

I have added these reflections to answer the possible objections of those who may contend that, in my discussion, I should have included the heat-producing and force-producing power of non-nitrogenized alimentary substances.

THE END.

LIGHT:

A SERIES OF SIMPLE, ENTERTAINING, AND INEXPENSIVE EXPERIMENTS IN THE PHENOMENA OF LIGHT, FOR THE USE OF STUDENTS OF EVERY AGE.

By ALFRED M. MAYER and CHARLES BARNARD.

Price, $1.00.

From the New York Evening Post.

"A singularly excellent little hand-book for the use of teachers, parents, and children. The book is admirable both in design and execution. The experiments for which it provides are so simple that an intelligent boy or girl can easily make them, and so beautiful and interesting that even the youngest children must enjoy the exhibition, while the whole cost of all the apparatus needed is only $12.40. The experiments here described are abundantly worth all that they cost in money and time in any family where there are boys and girls to be entertained."

From the New York Scientific American.

"The experiments are capitally selected, and equally as well described. The book is conspicuously free from the multiplicity of confusing directions with which works of the kind too often abound. There is an abundance of excellent illustrations."

From the American Journal of Science and Arts.

"The experiments are for the most part new, and have the merit of combining precision in the methods with extreme simplicity and elegance of design. The aim of the authors has been to make their readers 'experimenters, strict reasoners, and exact observers,' and for the attainment of this end the book is admirably adapted. Its value is further enhanced by the numerous carefully-drawn cuts, which add greatly to its beauty."

From the Boston Globe.

"The volume seems well adapted to the needs of students who like to have their knowledge vitalized by experiment. The fact that nearly all the experiments described are new, and have been tested, is an additional recommendation of this handy volume. The illustrations add to its interest and value, and its simplicity, both of design and execution, will commend it to beginners and others seeking information on the subject."

From the Philadelphia Press.

"It supplies a large number of simple and entertaining experiments on the phenomena of light, that any one can perform with materials that may be found in any dwelling-house, or that may be bought for a small sum in any town or city. This actually is philosophy in sport, which thoughtful or ready minds can easily convert into science in earnest."

D. APPLETON & CO., 549 & 551 BROADWAY, NEW YORK.

THE
ANCIENT LIFE-HISTORY
OF
THE EARTH.

A COMPREHENSIVE OUTLINE OF THE PRINCIPLES AND LEADING FACTS OF PALÆONTOLOGICAL SCIENCE.

By H. ALLEYNE NICHOLSON, M. D., F. R. S. E., F. L. S., Etc.,

PROFESSOR OF NATURAL HISTORY IN THE UNIVERSITY OF ST. ANDREWS; AUTHOR OF "MANUAL OF ZOÖLOGY," ETC., ETC.

WITH NUMEROUS ILLUSTRATIONS.

1 vol., small 8vo.   408 pages. . . .   Cloth, $2.00.

The Quarterly Journal of Science.

"A work by a master in the science who understands the significance of every phenomenon which he records, and knows how to make it reveal its lessons. As regards its value there can scarcely exist two opinions. As a text-book of the historical phase of palæontology it will be indispensable to students, whether specially pursuing geology or biology, and without it no man who aspires even to an outline knowledge of natural science can deem his library complete."

Athenæum.

"The Professor of Natural History in the University of St. Andrews has, by his previous works on zoölogy and palæontology, so fully established his claim to be an exact thinker and a close reasoner, that scarcely any recommendation of ours can add to the interest with which all students in natural history will receive the present volume. It is, as its second title expresses it, a comprehensive outline of the principles and leading facts of palæontological science. Numerous woodcut illustrations very delicately executed, a copious glossary, and an admirable index, add much to the value of this volume."

Nature.

"There is no feature in which ordinary geological manuals in common use are more deficient than in the sketches they give of the leading characteristics of the animal and vegetable life of the successive periods which they describe. We regard the present volume, therefore, as dealing with a subject in connection with which the want of a competent text-book has long been a serious evil; and of the general accuracy and reliability of the information supplied by it we can speak in terms of high commendation. Prof. Nicholson has wisely availed himself to the fullest extent of woodcut illustrations in aid of his descriptions of fossil specimens. In his discussion of the characters distinguishing the flora and fauna of each of the great geological periods, he is clear in his representations and happy in his choice of typical forms. His work constitutes a popular exposition and summary of the facts of palæontology, suitably arranged, and it is well worthy to take its place among the useful manuals for which we are already indebted to its industrious author."

Daily Telegraph.

"This work ought to be a boon not only to professional students, but to the general reader. The many illustrations, which are carefully and tastefully engraved, vastly enhance its value."

D. APPLETON & CO., Publishers, 549 and 551 Broadway, N. Y.

www.ingramcontent.com/pod-product-compliance
Lightning Source LLC
Chambersburg PA
CBHW030436190426
43202CB00036B/1539